Computational Aspects of Cooperative Game Theory

Synthesis Lectures on Artificial Intelligence and Machine Learning

Editors
Ronald J. Brachman, *Yahoo Research*
William W. Cohen, *Carnegie Mellon University*
Thomas Dietterich, *Oregon State University*

Algorithms for Reinforcement Learning
Csaba Szepesvári
2010

Data Integration: The Relational Logic Approach
Michael Genesereth
2010

Markov Logic: An Interface Layer for Artificial Intelligence
Pedro Domingos and Daniel Lowd
2009

Introduction to Semi-Supervised Learning
Xiaojin Zhu and Andrew B. Goldberg
2009

Action Programming Languages
Michael Thielscher
2008

Representation Discovery using Harmonic Analysis
Sridhar Mahadevan
2008

Essentials of Game Theory: A Concise Multidisciplinary Introduction
Kevin Leyton-Brown and Yoav Shoham
2008

A Concise Introduction to Multiagent Systems and Distributed Artificial Intelligence
Nikos Vlassis
2007

Intelligent Autonomous Robotics: A Robot Soccer Case Study
Peter Stone
2007

Computational Aspects of Cooperative Game Theory

Georgios Chalkiadakis, Edith Elkind, and Michael Wooldridge

ISBN: 978-3-031-00430-8   paperback
ISBN: 978-3-031-01558-8   ebook

DOI 10.1007/978-3-031-01558-8

A Publication in the Springer series
SYNTHESIS LECTURES ON ARTIFICIAL INTELLIGENCE AND MACHINE LEARNING

Lecture #16
Series Editors: Ronald J. Brachman, *Yahoo! Research*
                William W. Cohen, *Carnegie Mellon University*
                Thomas Dietterich, *Oregon State University*
Series ISSN
Synthesis Lectures on Artificial Intelligence and Machine Learning
Print 1939-4608   Electronic 1939-4616

Computational Aspects of Cooperative Game Theory

Georgios Chalkiadakis
Technical University of Crete, Greece

Edith Elkind
Nanyang Technological University, Singapore

Michael Wooldridge
University of Liverpool, United Kingdom

SYNTHESIS LECTURES ON ARTIFICIAL INTELLIGENCE AND MACHINE LEARNING #16

ABSTRACT

Cooperative game theory is a branch of (micro-)economics that studies the behavior of self-interested agents in strategic settings where binding agreements among agents are possible. Our aim in this book is to present a survey of work on the computational aspects of cooperative game theory. We begin by formally defining transferable utility games in characteristic function form, and introducing key solution concepts such as the core and the Shapley value. We then discuss two major issues that arise when considering such games from a computational perspective: identifying *compact representations* for games, and the closely related problem of *efficiently computing solution concepts* for games. We survey several formalisms for cooperative games that have been proposed in the literature, including, for example, cooperative games defined on networks, as well as general compact representation schemes such as MC-nets and skill games. As a detailed case study, we consider weighted voting games: a widely-used and practically important class of cooperative games that inherently have a natural compact representation. We investigate the complexity of solution concepts for such games, and generalizations of them. We briefly discuss games with non-transferable utility and partition function games. We then overview algorithms for identifying welfare-maximizing coalition structures and methods used by rational agents to form coalitions (even under uncertainty), including bargaining algorithms. We conclude by considering some developing topics, applications, and future research directions.

KEYWORDS

game theory, cooperative games, coalitional games, representations, computational complexity, solution concepts, core, Shapley value, coalition formation, coalition structure generation

This manuscript was a pleasure to discover, and a pleasure to read—a broad, but succinct, overview of work in computational cooperative game theory. I will certainly use this text with my own students, both within courses and to provide comprehensive background for students in my research group. The authors have made a substantial contribution to the multiagent systems and algorithmic game theory communities.

– Professor Jeffrey S. Rosenschein
   The Hebrew University of Jerusalem, Israel

With the advent of the internet, the computational aspects of cooperative game theory are ever more relevant. This unique and timely book by Chalkiadakis, Elkind, and Wooldridge gives a concise and comprehensive survey of the subject, and serves at the same time as a one-stop introduction to cooperative game theory.

– Professor Bernhard von Stengel
   London School of Economics, UK

In recent years, research on the computational aspects of cooperative game theory has made tremendous progress, but previous textbooks have not included more than a short introduction to this important topic. I am excited by the thorough treatment in this new book, whose authors have been and continue to be at the very forefront of this research. Newcomers to the area are well advised to read this book carefully and cover to cover.

– Professor Vincent Conitzer
   Duke University, USA

Cooperative game theory has proved to be a fertile source of challenges and inspiration for computer scientists. This book will be an essential companion for everyone wanting to explore the computational aspects of cooperative game theory.

– Prof Makoto Yokoo
   Kyushu University, Japan

An excellent treatise on algorithms and complexity for cooperative games. It navigates through the maze of cooperative solution concepts to the very frontiers of algorithmic game theory research. The last chapter in particular will be enormously valuable for graduate students and young researchers looking for research topics.

– Professor Xiaotie Deng
   University of Liverpool, UK

Georgios:

> *for my godmother, Katerina;*
> *and my godchildren, Filippos, Antonios, and Maria-Sofia.*

Edith:

> *to the memory of my grandfather Gersh.*

Michael:

> *for Lily May and Thomas Llewelyn.*

Contents

Preface

Over the past decade, there has been an enormous growth of interest in issues at the intersection of game theory and computer science. This phenomenon has been driven in part by the Internet juggernaut, which so spectacularly arrived in the public consciousness in the late 1990s. On the one hand, the Internet can be understood in purely computational terms: it is a massive, highly dynamic network of computational nodes, exchanging data across physical and wireless connections using carefully defined protocols and procedures. However, a purely computational view of the Internet surely misses aspects that are crucial to understanding it. The Internet is not populated by altruistic or even indifferent entities. The entities that make decisions and act on the Internet are *self-interested*. They have preferences, goals, and desires, and broadly speaking, they will act to bring about their preferences, goals, and desires as best they can. A traditional computer science analysis has nothing whatsoever to say about these *economic* aspects of the Internet. This observation motivated researchers to apply techniques from economics, and in particular, techniques from the branch of micro-economics called game theory, to the analysis of networked computer systems like the Internet. Research in this domain has come primarily from two communities: the *multiagent systems* community (which originated from research in artificial intelligence) [240, 261], and the *algorithmic game theory* community [195] (which originated from research in theoretical computer science). These two computer science communities have embraced techniques from game theory, and have made considerable progress in understanding both how game-theoretic techniques can be usefully applied in computational settings, and how computational techniques can be brought to bear on game-theoretic problems.

Our aim in the present book is to introduce the reader to research in the computational aspects of *cooperative* game theory. This is an important and extremely active area of contemporary research, of interest to both the multiagent systems and algorithmic game theory communities. Crudely, cooperative game theory (as opposed to non-cooperative game theory) is concerned with settings in which groups of agents can agree to cooperate with each other, for example, by signing a mutually binding contract. Binding agreements may enable the agents to implement solutions that are not considered possible in non-cooperative game theory.

Research in the computational aspects of cooperative game theory has two complementary aspects. On the one hand, we study the application of computational techniques to problems in cooperative game theory: how to compute outcomes of a game, for example. On the other hand, we might consider how concepts and techniques from cooperative game theory can be applied to computational problems. For example, as we will discuss, techniques from cooperative game theory can be usefully applied to the analysis of computer networks.

READERSHIP

Our book is a graduate-level text, and we expect the readership to consist primarily of graduate students and researchers who want to gain an understanding of the main issues and techniques in the field. We expect most readers to have a computer science background, but we hope the text is of interest and value to researchers in economics and game theory who want to gain an understanding of what computer scientists are doing with their ideas. If we have done our job well, then after reading this book, you should have a fighting chance of understanding most of the research papers that we cite in the book.

PRE-REQUISITES

We expect the readers of our book to be predominantly computer scientists and artificial intelligence researchers, with little or no background in game theory. For these readers, we provide in chapter 2 a complete (if somewhat terse) summary of the necessary definitions and concepts from cooperative game theory that are used throughout the remainder of the book.

Readers from the game theory community will find no surprises in our treatment of their subject, except that in our presentation, coalition structures play a more prominent role than is usually the case in the game theory literature. Specifically, we find it natural to introduce coalition structures from the start, defining the outcome of a cooperative game to be a coalition structure together with a payoff vector. Our reasoning here is that this helps in particular to clarify the story around superadditivity, how superadditivity implies the formation of the grand coalition, and solution concepts that are conventionally formulated with respect to the grand coalition (e.g., the Shapley value).

A key concern of our book is, of course, computation, and we make extensive use of techniques from the field of computational complexity in the analysis of cooperative games. A survey of the relevant techniques and concepts from this area is beyond the scope of the present book, and for this we refer the reader to [122, 203].

Throughout the book, we assume a modest level of mathematical ability: the math content is primarily discrete mathematics of the level and type that is usually taught in undergraduate computer science degrees.

WEB SITE

The following web site provides some links to relevant web resources, downloadable video lectures to accompany the book, and a complete set of lecture slides:

<code>http://web.spms.ntu.edu.sg/~eelkind/coopbook/</code>

Georgios Chalkiadakis, Edith Elkind, and Michael Wooldridge
October 2011

Acknowledgments

We are grateful to Mike Morgan for suggesting this volume, and for charming us into agreeing to write it. We thank Haris Aziz, Yoram Bachrach, Craig Boutilier, Paul E. Dunne, Leslie Ann Goldberg, Paul W. Goldberg, Nick Jennings, Ramachandra Kota, Vangelis Markakis, Dima Pasechnik, Mike Paterson, Maria Polukarov, Valentin Robu, Alex Rogers, Alex Skopalik, and Yair Zick for permitting us to adapt material from papers we have co-authored with them. The bulk of the material in this book is based on tutorials that were presented at ECAI-2008 (Patras, Greece), AAMAS-2009 (Budapest, Hungary), AAMAS-2010 (Toronto, Canada), AAAI-2010 (Atlanta, Georgia), AAMAS-2011 (Taipei, Taiwan), and IJCAI-2011 (Barcelona, Catalonia, Spain): we are grateful to the students who attended these tutorials for their questions, corrections, suggestions, and enthusiastic feedback. We would also like to thank Stéphane Airiau, Haris Aziz, Piotr Faliszewski, Felix Fischer, Martin Gairing, Reshef Meir, Dima Pasechnik, Tuomas Sandholm, Bernhard von Stengel, Yair Zick, and the anonymous reviewers arranged by the publisher, who carefully read a preliminary draft of the book and gave us detailed and enormously helpful feedback. It goes without saying that any remaining errors and omissions are entirely the responsibility of the authors. Edith Elkind gratefully acknowledges research fellowship RF2009-08 "Algorithmic aspects of coalitional games" from the National Research Foundation (Singapore).

PERSONAL PRONOUNS

We find it inappropriate to refer to agents solely as "he" or solely as "she", and calling agents "it" seems just plain ugly. So, we interleave usages of "he" and "she" (semi-)randomly.

Georgios Chalkiadakis, Edith Elkind, and Michael Wooldridge
October 2011

Summary of Key Notation

G A game. (Chapter 1)

Γ A class of games. (Chapter 1)

Ω The outcomes of a game; where $G \in \Gamma$, then Ω_G is the set of possible outcomes of G. (Chapter 1)

$\sigma : \Gamma \to 2^{\Omega}$ A solution concept: for every game $G \in \Gamma$, identifies a set of outcomes $\sigma(G) \subseteq \Omega_G$ that are intuitively the "rational" outcomes of the game according to σ. (Chapter 1)

$N = \{1, \ldots, n\}$ The players in a game. (Chapter 2)

$C, C', C^1, C_1 \ldots$ Coalitions, i.e., subsets of players N. (Chapter 2)

$v : 2^N \to \mathbb{R}$ The characteristic function of a game: assigns a real number to every possible coalition, indicating the value that this coalition could obtain if they cooperated. (Chapter 2)

$G = (N, v)$ A characteristic function game; the basic model of cooperative games we use throughout the book. (Chapter 2)

$CS = \{C^1, \ldots, C^k\}$ A coalition structure: a partition of N into mutually disjoint coalitions, i.e., $C^1 \cup \cdots \cup C^k = N$, and $C^i \cap C^j = \emptyset$ for $i \neq j$. (Chapter 2)

$\mathcal{CS}_N$ The set of all coalition structures over N. (Chapter 2)

$v(CS)$ The social welfare of coalition structure CS: $v(CS) = \sum_{C \in CS} v(C)$. (Chapter 2)

$\mathbf{x} = (x_1, \ldots, x_n)$ A payoff vector: a distribution of payoffs to players in a game; x_i is the payoff received by player i. (Chapter 2)

$x(C)$ The total payoff of a coalition $C \subseteq N$ under $\mathbf{x}$, i.e. $x(C) = \sum_{i \in C} x_i$. (Chapter 2)

$\mathcal{I}(CS)$ The set of all imputations for a coalition structure CS. (Chapter 2)

$(CS, \mathbf{x})$ An outcome of a cooperative game: a coalition structure together with a payoff vector. (Chapter 2)

G^* The superadditive cover of game G. (Chapter 2)

Π_N The set of all permutations (possible orderings) of players N. (Chapter 2)

π An element of Π_N, i.e., a permutation of the players N. (Chapter 2)

$S_\pi(i)$ The set of players preceding player $i \in N$ in the ordering π. For example if $\pi = (3, 1, 2)$ then $S_\pi(1) = \{3\}$, $S_\pi(2) = \{1, 3\}$. (Chapter 2)

$\Delta_\pi^G(i)$ The marginal contribution that player i makes to the set of players preceding him in the ordering π in game G. (Chapter 2)

$\varphi_i(G)$ The Shapley value of player $i \in N$ in game G. (Chapter 2)

$\beta_i(G)$ The Banzhaf index of player $i \in N$ in game G. (Chapter 2)

$\eta_i(G)$ The normalized Banzhaf index of player $i \in N$ in game G. (Chapter 2)

$\mathcal{C}(G)$ The core of G. (Chapter 2)

$d(\mathbf{x}, C)$ The deficit of C with respect to $\mathbf{x}$, i.e., the amount that C could gain by deviating from $\mathbf{x}$: $d(\mathbf{x}, C) = v(C) - x(C)$. (Chapter 2)

$\mathbf{d}(\mathbf{x})$ The deficit vector of $\mathbf{x}$. (Chapter 2)

$\mathcal{N}(G)$ The nucleolus of game G. (Chapter 2)

$S_{i,j}(\mathbf{x})$ The surplus of player i over player j with respect to payoff vector $\mathbf{x}$. (Chapter 2)

$\mathcal{K}(G)$ The kernel of game G. (Chapter 2)

$\mathcal{B}(G)$ The bargaining set of game G. (Chapter 2)

$G = [N; \mathbf{w}; q]$ A weighted voting game with players N, weights $\mathbf{w} = (w_1, \ldots, w_n) \in \mathbb{R}^n$, and quota $q \in \mathbb{R}$. (Chapter 4)

$w(C)$ The total weight of coalition C in a weighted voting game: $w(C) = \sum_{i \in C} w_i$. (Chapter 4)

$w_{\max}$ The largest weight of any player in a given weighted voting game. (Chapter 4)

$\dim(G)$ The dimension of vector weighted voting game G. (Chapter 4)

$\Lambda = \{\lambda_1, \ldots\}$ The set of choices in an NTU game. (Chapter 5)

$\succeq_i \subseteq \Lambda \times \Lambda$ Preference relation for player $i \in N$ over choices Λ; the relation is required to be complete, reflexive, and transitive. (Chapter 5)

$\mathbf{c} = (\lambda_1, \ldots, \lambda_k)$ A choice vector: a tuple of choices, one for each coalition in a coalition structure. (Chapter 5)

$(CS, \mathbf{c})$ An outcome for an NTU game: a coalition structure together with a vector of choices, one for each coalition in the coalition structure, where the choices in the choice vector must be feasible for the corresponding coalition. (Chapter 5)

$v : 2^N \to 2^\Lambda$ Characteristic function for an NTU game: maps each coalition to the set of choices available to that coalition. (Chapter 5)

$G = (N, \Lambda, v, \succeq_1, \ldots, \succeq_n)$ An NTU game. (Chapter 5)

θ A goal in a qualitative game. (Chapter 5)

Θ The set of all goals in a qualitative game. (Chapter 5)

Θ_i The goal set of agent $i \in N$ in a qualitative game. (Chapter 5)

Φ A set of Boolean variables. (Chapter 5)

Φ_i The set of Boolean variables under the control of player $i \in N$ in a Boolean game. (Chapter 5)

(C, CS) An embedded coalition, i.e., a coalition structure $CS \in \mathcal{CS}_N$ together with a coalition $C \in CS$ from this structure. (Chapter 5)

E_N The set of all embedded coalitions over N. (Chapter 5)

$u : E_N \to \mathbb{R}$ The partition function game analogue of a characteristic function: gives the value of every embedded coalition. (Chapter 5)

$CS^*(G)$ The socially optimal coalition structure of G. (Chapter 6)

CHAPTER 1

Introduction

This book presents a survey of research that has been carried out by two closely related research communities: the *multiagent systems* community and the *algorithmic game theory* community. The multiagent systems community studies interactions between *self-interested computational entities*, called *agents*, which are typically assumed to be acting on behalf of human users or owners [240, 261]. A key aim of the multiagent systems research area is to build computer systems that can effectively and autonomously cooperate in order to achieve goals delegated by users. The *algorithmic game theory* research community has very closely related concerns [195]. Algorithmic game theory is fundamentally concerned with algorithms in strategic settings, where those providing the inputs to algorithms are assumed to be acting in their own best interests.

Given the overall aims of the multiagent systems community and the algorithmic game theory community, it is very natural to ask what mathematical models are available that can help us to understand systems containing multiple self-interested entities, and the interactions that might occur within then. *Game theory* is a branch of (micro-)economics that is largely concerned with the theory of decision making in environments populated by self-interested entities [47, 174, 191, 200]. Game theory uses mathematical models known as games to capture the key attributes of scenarios in which self-interested agents interact. Before proceeding, a word of caution: the term "game" in this book is used in a technical sense—the technical sense of game theory. We do not mean games in the everyday recreational sense (chess, checkers, poker, ...), although the use of the term "game" (and much of the associated theory) was originally derived from the study of recreational games such as poker.

A key concern in game theory is to try to understand what counts as a *rational outcome* of a game. To this end, game theorists have developed a number of *solution concepts*, which for every game identify some subset of the possible outcomes of the game. If we think of a solution concept as capturing some notion of rationality, then the outcomes of a game identified in this way will represent the rational outcomes of the game[1]. Common problems with solution concepts are that they may not in general guarantee the existence of a rational outcome, and they may not guarantee that a rational outcome is unique. Such problems have led to the development of a range of different solution concepts, capturing different notions of rationality.

---

[1]There is a substantial ongoing debate about what solution concepts *mean*. Under a *normative* interpretation, a solution concept characterizes what a rational player *should do*, while under a *descriptive* interpretation, a solution concept characterizes what players *actually do*. A standard criticism of many solution concepts is that they are not descriptive, in that they fail to predict what people will actually do. This debate is well beyond the scope of the present book, and we do nothing more in this book than acknowledge the debate here.

A key distinction in game theory is made between games that are *non-cooperative* and games that are *cooperative*. Non-cooperative games represent arguably the best-studied class of games, and most examples of recreational games (chess, checkers, …) are non-cooperative. Before proceeding any further, let us take some time to understand the distinction between cooperative and non-cooperative games.

1.1 WHY ARE NON-COOPERATIVE GAMES NON-COOPERATIVE?

We consider an example of a game that is not cooperative, although it very much seems like it *should* be cooperative. The game is called the *Prisoner's Dilemma*, and it is perhaps the most famous example of a non-cooperative game [17]. The Prisoner's Dilemma is traditionally described by the following scenario:

Two men, *A* and *B*, are collectively charged with a crime and held in separate cells, *with no way of meeting or communicating, and no way to make binding agreements.* They are told that:

- if one confesses and the other does not, the confessor will be freed, and the other will be jailed for three years;

- if both confess, then each will be jailed for two years.

Both prisoners know that if neither confesses, then they will each be jailed for one year.

Intuitively, the prisoners have to decide whether to *cooperate* (keep their mouths shut—not confess to the crime) and *not cooperate* (confess). How should a player choose rationally between these two strategies? To answer this question, consider the following line of reasoning, from the point of view of player *A*:

Suppose *B* confesses; then, if I confess, my prison term would be two years, and if I keep quiet, it would be three years. Thus, my best choice would be to confess. But suppose *B* keeps quiet; then, if I confess, I would walk free, and if I keep quiet, I would spend a year in jail. Again, my best choice would be to confess. So, no matter what *B* does, my best course of action is to confess.

Of course, the Prisoner's Dilemma is *symmetric*: player *B* will reason in the same way about player *A*, and conclude that his best choice is also to confess. The upshot is that they *both* confess, and the overall outcome of the game is that both prisoners will serve 2 years in jail.

But common sense seems to tell us that this outcome is not the best that could be done…why don't both players *cooperate* by keeping quiet? This would lead to the outcome where both players would serve 1 year in jail. Such mutual cooperation would be *strictly* preferred over mutual confession by *both* players.

What we have described here is in fact a very strong solution concept known as *dominant strategy equilibrium*. Dominant strategy equilibria do not always exist in games, but where they do, it is very hard to imagine any other outcome occurring through rational choice. The Prisoner's Dilemma is a dilemma precisely because the unique rational outcome, according to dominant strategy equilibrium, is strictly worse for both players than another outcome. The rational outcome seems to be sub-optimal for *both* players.

So, why doesn't cooperation—in the sense of both players keeping quiet—occur rationally in the Prisoner's Dilemma? Put simply, cooperation cannot occur in the Prisoner's Dilemma *because the conditions required for cooperation are not present* [50]. And in particular:

binding agreements are not possible.

In fact, we designed this assumption into the structure of the game: we explicitly stated, as a background assumption, that the players had no way of meeting or communicating. Because binding agreements are not possible, the players in the game cannot trust one another: they must make their decisions based solely on the information they have about the possible choices and corresponding utilities, will try to maximize their own utility (minimize the amount of time spent in jail), and must assume that all other players are trying to do the same thing. Notice that if binding agreements were possible in the Prisoner's Dilemma (i.e., the prisoners were able to get together beforehand and make irrevocable commitments to some course of action), then there would indeed be no dilemma at all: they could both make a binding agreement to cooperate by keeping quiet, and both would do better than the dominant strategy equilibrium of mutual confession.

In many real-world situations, the assumption that binding agreements are not possible simply does not hold. In the real world, we can use contracts and other legal arrangements to make binding agreements with others. Put simply, cooperative game theory studies games in which binding agreements *are* possible:

> [A game] is cooperative if the players can make binding agreements about the distribution of payoffs or the choice of strategies, even if these agreements are *not* specified or implied by the rules of the game. [207, p.1]

It seems to us that most research in game theory has focused on non-cooperative games like the Prisoner's Dilemma, and solution concepts for non-cooperative games, such as the ubiquitous Nash equilibrium [200]. Comparatively, cooperative game theory seems to have attracted less attention (although we emphasize that cooperative game theory is in itself a large and well-established research field). This is perhaps surprising, because binding agreements—in the form of contracts—are what makes commerce, and the global economy, possible:

> Binding agreements are prevalent in economics. Indeed, almost every [...] seller–buyer transaction is binding. [207, p.1]

So, in this book, we will focus on games in which binding agreements are possible. Notice that game theory itself does not consider the form that binding agreements take, and we also do not consider this issue. We simply assume that some mechanism for forming binding agreements exists.

1.2   COMPUTATIONAL PROBLEMS IN GAME THEORY

Our book is distinguished from more conventional game theory texts by the fact that we consider *computational* aspects of cooperative game theory. So let us say a few words at this stage about what computational problems we might consider with respect to games. We introduce some light notation to help formulate these problems.

First, let Γ be a class of games. For our present purposes, it does not matter exactly what form the games $G \in \Gamma$ take. Associated with Γ is a set Ω of *outcomes*. Where $G \in \Gamma$ is a specific game, let Ω_G denote the possible outcomes of G. For example, taking G to be the Prisoner's Dilemma, as described above, there are four possible outcomes of the game: (confess, confess), (confess, keep quiet), (keep quiet, confess), and (keep quiet, keep quiet). Each of these four outcomes assigns a utility to each player.

Given this notation, we can model a solution concept σ for a class of games Γ with outcomes Ω as a function:

$$\sigma : \Gamma \to 2^{\Omega},$$

where σ is required to satisfy the property that $\sigma(G) \subseteq \Omega_G$. Thus, for every game $G \in \Gamma$, the solution concept σ identifies some subset of the possible outcomes Ω_G of G. If σ is intended to capture some notion of rational outcome, then $\sigma(G)$ would be the possible outcomes of the game if the players employed this notion of rational choice. Again, referring back to the Prisoner's Dilemma, the solution concept of dominant strategy equilibrium identifies a single outcome as the solution of the game: (confess, confess).

Given a class of games Γ with outcomes Ω, and a solution concept $\sigma : \Gamma \to 2^{\Omega}$, a number of problems suggest themselves:

Non-emptiness: Given $G \in \Gamma$, is it the case that $\sigma(G) \neq \emptyset$? Thus, non-emptiness simply asks whether the game has any outcome that is rational according to the solution concept σ.

Membership: Given $G \in \Gamma$ and $\omega \in \Omega_G$, is it the case that $\omega \in \sigma(G)$? This problem basically asks whether a given candidate outcome is rational according to solution concept σ.

Computation: Given $G \in \Gamma$, output some ω such that $\omega \in \sigma(G)$. In this problem, we actually want to compute a rational outcome of the game.

If we consider these as computational problems, then the first two are decision problems; the final problem is a function problem. Now, solution concepts σ are typically *strategic optimization* problems, in the sense that the solutions they propose capture some notion of optimality in a strategic setting. Combinatorial optimization problems are of course very important and well-studied in computer science, and they are in general computationally hard to solve [203, 204]. It should therefore come as no surprise that the computational problems discussed above are often computationally hard—NP-hard or worse. Thus, a key issue in the multiagent systems and algorithmic game theory communities has been the development of techniques for efficiently computing solution concepts, and on the development of computationally tractable solution concepts.

With all of the questions listed above, an important issue is that of *representation*. A standard consideration in computational complexity and algorithm analysis is that of how the *inputs* to a computational problem are encoded or represented. In the problems listed above, the inputs are games and the outcomes for games. Thus, to formally analyze the problems listed above in computational terms, we need some way of representing or encoding games and their outcomes. A rule of thumb in computational analysis is that the more concise or expressive a representation scheme used to encode inputs is, the harder will be the associated computational problems. This issue is emphasized rather dramatically when we look at the computational aspects of cooperative game theory, because, as we will see soon, the "obvious" representations for cooperative games are unrealistic and completely unusable in practice: they lead to representations of games that are *of size exponential in the number of players*. Thus, if we want to compute with models of cooperative games, we have no option but to look for concise, succinct representations. But, as we just mentioned, succinct representations tend to lead to higher computational complexity. So, as the reader will understand after reading this book, much research in the computational aspects of cooperative game theory is about finding representations that are compact and expressive enough to be useful, yet admit efficient algorithms for computing answers to the problems listed above.

1.3   THE REMAINDER OF THIS BOOK

The remainder of this book is structured as follows.

- We begin, in the following chapter, by introducing the relevant definitions and technical concepts from cooperative game theory that are used throughout the remainder of the book. Specifically, we introduce transferable utility characteristic function games, and the main solution concepts for such games—the Shapley value, the core, the kernel, the nucleolus, and some variations on these.

- In chapter 3, we introduce a variety of models and representations for cooperative games, and consider algorithms for computing cooperative solution concepts for these representations. The two key themes in this chapter are *succinct* representations, and the *complexity* of solution concepts for these representations.

- In chapter 4, we focus on *simple games*, and more specifically, *weighted voting games*—one of the most important classes of cooperative games, frequently found in the real world.

- In chapter 5, we briefly discuss *NTU games* and *partition function games*. Both of these classes of games generalize transferable utility characteristic function games, albeit in different directions: In an NTU game, the benefit that an individual player obtains cannot be transferred to other players in a coalition, and in a partition function game a coalition's payoff may depend on how the rest of the agents are split into coalitions.

- In chapter 6, we discuss the problem of *coalition structure formation*: how to optimally partition a collection of agents into disjoint teams—both from the point of view of a central designer, and that of the individual agents.

- In chapter 7, we discuss some advanced topics (e.g., the links between mechanism design and cooperative game theory) and some applications of cooperative game theory.

1.4   FURTHER READING

For a detailed introduction to contemporary research issues in algorithmic game theory, see [195]. For an undergraduate introduction to multiagent systems, see [261], and for a detailed introduction to the game-theoretic aspects of multiagent systems research, see [240]. There are many excellent textbooks on game theory, and most of these have some coverage of cooperative game theory; Osborne and Rubinstein's *A Course in Game Theory* is a good example [200]. Other good textbooks on game theory and micro-economics with coverage of cooperative game theory include [174, 191]. Binmore's *Very Short Introduction to Game Theory* is a lucid and enjoyable informal introduction to the scope and key concepts of game theory [50]. There are fewer books that focus exclusively on cooperative game theory [61, 82, 100]; a good introduction to the mathematics of cooperative games is [207]. Simple games are discussed in detail in [251].

CHAPTER 2

Basic Concepts

The goal of this chapter is to introduce the basic notions of cooperative game theory, which are then used in the remainder of the book. We will begin by formally defining *characteristic function games* and some of their subclasses, and then present the standard solution concepts for such games.

2.1 CHARACTERISTIC FUNCTION GAMES

A game in the sense of game theory is an abstract mathematical model of a scenario in which self-interested agents interact. It is abstract in the sense that irrelevant detail is omitted: the game aims to capture only those features of the scenario that are relevant to the decisions that must be made by players within the game. The form of games that we now introduce is the most basic and widely-studied model of cooperative games.

Throughout this book, we will consider games that are populated by a non-empty set $N = \{1, \ldots, n\}$ of *agents*: the *players* of the game. A *coalition* is simply a subset of the players N. We will use $C, C', C^1, C_1 \ldots$ to denote coalitions. The *grand coalition* is the set N of all players. Note that in everyday use, the term "coalition" usually implies a group of agents who have some kind of commitment to common action. This is not the sense in which we use the term: in this book, a coalition is simply a subset of players.

Definition 2.1   A *characteristic function game* G is given by a pair (N, v), where $N = \{1, \ldots, n\}$ is a finite, non-empty set of *agents* and $v : 2^N \to \mathbb{R}$ is a *characteristic function*, which maps each coalition $C \subseteq N$ to a real number $v(C)$. The number $v(C)$ is usually referred to as the *value* of the coalition C.

Notice that this model gives us no indication of how the characteristic function v should be derived for any given cooperative scenario. As we will see, there are many possible interpretations for the characteristic function. In addition, note that characteristic function games assign the value of a coalition to the coalition as a whole, and not to its individual members. That is, the characteristic function game model does not dictate how the coalitional value $v(C)$ should be divided amongst the members of C. In fact, the question of how to divide coalitional value is a fundamental research topic in cooperative game theory, and later in this chapter, we will see some answers to this question, in the form of solution concepts such as the Shapley value and the nucleolus. As an aside, notice that an implicit assumption in characteristic function games is that the coalitional value $v(C)$ can be divided amongst the members of C in any way that the members of C choose. Formally, games with this property are said to be *transferable utility games* (TU games). Most of the games we will study in this book are TU games, but in chapter 5 we will motivate and discuss some games in which utility is not transferable in this way.

Now, recall from chapter 1 that if we want to study games from a computational point of view, we need to consider how such games are represented or encoded as inputs to a program. The obvious representation of a characteristic function game suggested by Definition 2.1 is to explicitly list every coalition $C \subseteq N$ together with the associated value $v(C)$. However, it should be immediately obvious that this representation is of size $\Omega(2^n)$, i.e., exponential in the number of players in the game. It follows that this naive representation is not practical unless the number of players is very small. On the other hand, as argued by Megiddo [179], most real-life interactions that can be usefully modeled as cooperative games admit an encoding of size poly(n); such an encoding provides an *implicit* description of the characteristic function. Many such compact representations have been considered in the literature; we will survey some of them in chapters 3 and 4.

We will now present two examples of characteristic function games. In our first example, groups of children can pool their money to buy ice cream. While this scenario does not perhaps qualify as an important application of cooperative game theory, the same mathematical model can be used to describe any group of agents that can pool their resources to execute tasks with different payoffs and resource requirements.

Example 2.2   Charlie (C), Marcie (M), and Pattie (P) want to pool their savings to buy ice cream. Charlie has c dollars, Marcie has m dollars, Pattie has p dollars, and the ice cream tubs come in three different sizes: 500g, which costs \$7; 750g, which costs \$9; and 1000g, which costs \$11. The children value ice cream, and assign no utility to money. Thus, the value of each coalition is determined by how much ice cream it can buy.

This situation corresponds to a characteristic function game with the set of players $N = \{C, M, P\}$. For $c = 3, m = 4, p = 5$, its characteristic function v is given by the following table:

S	$\emptyset$	$\{C\}$	$\{M\}$	$\{P\}$	$\{C, M\}$	$\{C, P\}$	$\{M, P\}$	$\{C, M, P\}$
$v(S)$	0	0	0	0	500	500	750	1000

For $c = 8, m = 8, p = 1$, the characteristic function v is given by

S	$\emptyset$	$\{C\}$	$\{M\}$	$\{P\}$	$\{C, M\}$	$\{C, P\}$	$\{M, P\}$	$\{C, M, P\}$
$v(S)$	0	500	500	0	1250	750	750	1250

Our second example models decision-making in voting bodies. The game we describe belongs to the class of *weighted voting games*; these games are studied in detail in chapter 4.

Example 2.3   A fictional country X has a 101-member parliament, where each representative belongs to one of the four parties: Liberal (L), Moderate (M), Conservative (C), or Green (G). The Liberal party has 40 representatives, the Moderate party has 22 representatives, the Conservative party has 30 representatives, and the Green party has 9 representatives. The parliament needs to decide how to allocate \$1bn of discretionary spending, and each party has its own preferred way of using this money. The decision is made by a simple majority vote, and we assume that all representatives vote along the party lines. Parties can form coalitions; a coalition has value $1bn if it can win the budget vote no matter what the other parties do, and value 0 otherwise.

This situation can be modeled as a 4-player characteristic function game, where the set of players is $N = \{L, M, C, G\}$ and the characteristic function v is given by

$$v(S) = \begin{cases} 0 & \text{if } |S| \leq 1, \text{ or } |S| = 2 \text{ and } G \in S \\ 10^9 & \text{otherwise.} \end{cases}$$

2.1.0.1 Standard Assumptions

It is usually assumed that the value of the empty coalition $\emptyset$ is 0; we will make this assumption throughout the book. Moreover, it is often the case that the value of each coalition is non-negative (i.e., agents form coalitions to make a profit), or else that the value of each coalition is non-positive (i.e., agents form coalitions to share costs). Throughout this book, we will mostly focus on the former scenario, i.e., we assume that $v(C) \geq 0$ for all $C \subseteq N$. However, most of our definitions and results can be easily adapted to the latter scenario.

2.1.1 OUTCOMES

Recall that a key concern in game theory is to try to understand what the outcomes of a game will be. In order to answer this question for cooperative games, we must first consider what form the outcomes of a cooperative game will take. An outcome of a characteristic function game consists of two parts:

- a partition of players into coalitions, called a *coalition structure*; and

- a *payoff vector*, which distributes the value of each coalition among its members.

We will now describe both of these parts more formally.

Definition 2.4 Given a characteristic function game $G = (N, v)$, a *coalition structure* over N is a collection of non-empty subsets $CS = \{C^1, \ldots, C^k\}$ such that

- $\bigcup_{j=1}^{k} C^j = N$, and

- $C^i \cap C^j = \emptyset$ for any $i, j \in \{1, \ldots, k\}$ such that $i \neq j$.

A vector $\mathbf{x} = (x_1, \ldots, x_n) \in \mathbb{R}^n$ is a *payoff vector* for a coalition structure $CS = \{C^1, \ldots, C^k\}$ over $N = \{1, \ldots, n\}$ if

- $x_i \geq 0$ for all $i \in N$, and

- $\sum_{i \in C^j} x_i \leq v(C^j)$ for any $j \in \{1, \ldots, k\}$.

An *outcome* of G is a pair $(CS, \mathbf{x})$, where CS is a coalition structure over G and $\mathbf{x}$ is a payoff vector for CS. Given a payoff vector $\mathbf{x}$, we write $x(C)$ to denote the total payoff $\sum_{i \in C} x_i$ of a coalition $C \subseteq N$ under $\mathbf{x}$; we will use similar notation for other n-dimensional vectors throughout the book.

To understand the meaning of the conditions in Definition 2.4, let us consider them one by one. The first requirement in the definition of a coalition structure states that every player must appear in some coalition; the second says that a player cannot appear in more than one coalition. As for the payoff vector, the first condition states that every player receives a non-negative payoff. The second condition is a *feasibility* requirement, which states that the total amount paid out to a coalition cannot exceed the value of that coalition. We will often assume a slightly stronger *efficiency* requirement: a payoff vector $\mathbf{x}$ is *efficient* if *all* the payoff obtained by a coalition is distributed amongst coalition members, i.e., $\sum_{i \in C^j} x_i = v(C^j)$ for every $j \in \{1, \ldots, k\}$.

We will denote the space of all coalition structures over N by $\mathcal{CS}_N$. For example, suppose we have three agents, $N = \{1, 2, 3\}$; then there are seven possible non-empty coalitions:

$$\{1\}, \{2\}, \{3\}, \{1, 2\}, \{2, 3\}, \{3, 1\}, \{1, 2, 3\}$$

and five possible coalition structures:

$$\mathcal{CS}_{\{1,2,3\}} = \{\{\{1\}, \{2\}, \{3\}\}, \ \{\{1\}, \{2, 3\}\}, \ \{\{2\}, \{1, 3\}\}, \ \{\{3\}, \{1, 2\}\}, \ \{\{1, 2, 3\}\}\}.$$

Abusing notation, we will write $v(CS) = \sum_{C \in CS} v(C)$; the quantity $v(CS)$ is referred to as the *social welfare* of the coalition structure CS.

A payoff vector $\mathbf{x}$ for a coalition structure $CS \in \mathcal{CS}_N$ is said to be an *imputation* if it is efficient and moreover satisfies the *individual rationality* condition, i.e., $x_i \geq v(\{i\})$ for all $i \in N$. We denote by $\mathcal{I}(CS)$ the set of all imputations for a coalition structure CS; when $CS = \{N\}$, we write simply $\mathcal{I}(N)$. If a payoff vector is an imputation, each player weakly prefers being in the coalition structure to being on his own. Now, of course, players may still find it profitable to deviate *as a group*; we will discuss the issue of stability against group deviations in section 2.2. However, before we do that, let us consider a few important classes of characteristic function games, and discuss the relationships among them.

2.1.2   SUBCLASSES OF CHARACTERISTIC FUNCTION GAMES

We will now define four important subclasses of coalitional games: monotone games, superadditive games, convex games, and simple games.

2.1.2.1 Monotone Games

In many cases, adding an agent to an existing coalition can only increase the overall productivity of this coalition; games with this property are said to be *monotone*.

Definition 2.5   A characteristic function game $G = (N, v)$ is said to be *monotone* if it satisfies $v(C) \leq v(D)$ for every pair of coalitions $C, D \subseteq N$ such that $C \subseteq D$.

Clearly, the games described in Examples 2.2 and 2.3 are monotone—as are most other games considered in this book. However, it would be too restrictive to assume that coalitional games are always monotone. For instance, non-monotonicity may be caused by communication and coordination costs: while a new player may contribute to the total productivity of a coalition, he also needs to coordinate his actions with the other members of the coalition, and at some point the associated overheads may exceed the benefits derived from his participation. Alternatively, some players may strongly dislike each other and cannot work together productively, in which case adding a player to a coalition that contains his "enemy" may lower that coalition's value.

2.1.2.2 Superadditive Games

A stronger property that is also enjoyed by many practically useful games is *superadditivity*: in superadditive games, it is always profitable for two groups of players to join forces.

Definition 2.6   A characteristic function game $G = (N, v)$ is said to be *superadditive* if it satisfies $v(C \cup D) \geq v(C) + v(D)$ for every pair of disjoint coalitions $C, D \subseteq N$.

Since we have assumed that the value of each coalition is non-negative, superadditivity implies monotonicity: if a game $G = (N, v)$ is superadditive, and $C \subseteq D$, then $v(C) \leq v(D) - v(D \setminus C) \leq v(D)$. However, the converse is not necessarily true: consider, for instance, a game where the value of the characteristic function grows logarithmically with the coalition size, i.e., $v(C) = \log |C|$. In practice, non-superadditive games may arise as a result of anti-trust or anti-monopoly laws. In many countries such laws prohibit certain companies from working together in certain ways, making mergers unprofitable.

In superadditive games, there is no compelling reason for agents to form a coalition structure consisting of multiple coalitions: the agents can earn at least as much profit by working together within the grand coalition. Therefore, for superadditive games it is usually assumed that the agents form the grand coalition, i.e., the outcome of a superadditive game is of the form $(\{N\}, \mathbf{x})$ where $\mathbf{x}$ satisfies $\sum_{i \in N} x_i = v(N)$. Conventionally, N is omitted from the notation, i.e., an outcome of a superadditive game is simply a payoff vector for the grand coalition.

Any non-superadditive game can be transformed into a superadditive game by computing, for each coalition, the maximum amount this coalition can earn by splitting into subcoalitions. Formally, given a (non-superadditive) game $G = (N, v)$, we can define a new game $G^* = (N, v^*)$ by setting

$$v^*(C) = \max_{CS \in \mathcal{CS}_C} v(CS),$$

for every coalition $C \subseteq N$, where $\mathcal{CS}_C$ denotes the space of all coalition structures over C. The game G^* is called the *superadditive cover* of G; it is not hard to see that it is superadditive even if G is not. Intuitively, the value of a coalition C in G^* is the maximum amount that the players in C can make if they are free to choose their collaboration pattern.

2.1.2.3 Convex Games

The superadditivity property places a restriction on the behavior of the characteristic function v on disjoint coalitions. By placing a similar restriction on v's behavior on non-disjoint coalitions, we obtain the class of *convex* games.

Definition 2.7   A characteristic function v is said to be *supermodular* if it satisfies

$$v(C \cup D) + v(C \cap D) \geq v(C) + v(D)$$

for every pair of coalitions $C, D \subseteq N$. A game with a supermodular characteristic function is said to be *convex*.

Convex games have a very intuitive characterization in terms of players' marginal contributions: in a convex game, a player is more useful when she joins a bigger coalition.

Proposition 2.8   *A characteristic function game $G = (N, v)$ is convex if and only if for every pair of coalitions T, S such that $T \subset S$ and every player $i \in N \setminus S$ it holds that*

$$v(S \cup \{i\}) - v(S) \geq v(T \cup \{i\}) - v(T).$$

Proof. For the "only if" direction, assume that $G = (N, v)$ is convex, and consider two coalitions T, S such that $T \subset S \subset N$ and a player $i \in N \setminus S$. By setting $C = S$, $D = T \cup \{i\}$, we obtain

$$v(S \cup \{i\}) - v(S) = v(C \cup D) - v(C) \geq v(D) - v(C \cap D) = v(T \cup \{i\}) - v(T),$$

which is exactly what we need to prove. The "if" direction can be proved by induction on the size of $C \setminus D$.   □

Any convex game is necessarily superadditive: if a game $G = (N, v)$ is convex, and C and D are two disjoint subsets of N, then we have $v(C \cup D) \geq v(C) + v(D) - v(C \cap D) = v(C) + v(D)$ (here we use our assumption that $v(\emptyset) = 0$). However, the converse is not always true, as illustrated by the following example.

Example 2.9   Consider a game $G = (N, v)$, where $N = \{1, 2, 3\}$, and $v(C) = 1$ if $|C| \geq 2$ and $v(C) = 0$ otherwise; this game is known as the 3-*player majority game*. It is easy to check that this game is superadditive. On the other hand, for $C = \{1, 2\}$ and $D = \{2, 3\}$, we have $v(C) = v(D) = 1$, $v(C \cup D) = 1$, $v(C \cap D) = 0$.

2.1.2.4 Simple Games

Another well-studied class of coalitional games is that of *simple games*: a game $G = (N, v)$ is said to be *simple* if it is monotone and its characteristic function only takes values 0 and 1, i.e., $v(C) \in \{0, 1\}$ for any $C \subseteq N$. For instance, the game in Example 2.3 becomes a simple game if we rescale the payoffs so that $v(N) = 1$. In a simple game, coalitions of value 1 are said to be *winning*, and coalitions of value 0 are said to be *losing*. Such games can model situations where there is a task to be completed: a coalition is winning if and only if it can complete the task.

Note that simple games are superadditive only if the complement of each winning coalition is losing. Clearly, there exist simple games that are not superadditive. Nevertheless, it is usually assumed that the outcome of a simple game is a payoff vector for the grand coalition, just as in superadditive games.

2.2   SOLUTION CONCEPTS

Any partition of agents into coalitions and any payoff vector that respects this partition corresponds to an outcome of a characteristic function game. However, not all outcomes are equally desirable, or equally likely to occur. For instance, if all agents contribute equally to the value of a coalition, a payoff vector that allocates the entire payoff to just one of the agents is less appealing than the one that shares the profits equally among all agents. Similarly, an outcome that incentivizes all agents to work together is preferable to an outcome that some of the agents want to deviate from.

More broadly, one can evaluate the outcomes according to two sets of criteria: (1) *fairness*, i.e., how well each agent's payoff reflects his contribution, and (2) *stability*, i.e., what are the incentives for the agents to stay in the coalition structure. These two sets of criteria give rise to two families of solution concepts. We will now discuss each of them in turn.

2.2.1   SHAPLEY VALUE

The first solution concept we consider aims to capture the notion of fairness in characteristic function games. It is known as the *Shapley value* after its inventor Lloyd S. Shapley [231]. The Shapley value is a solution concept that is usually formulated with respect to the grand coalition: it defines a way of distributing the value $v(N)$ that could be obtained by the grand coalition. For this reason, it is often implicitly assumed that the games to which it is applied are superadditive (you will recall that in superadditive games, it makes sense for the grand coalition to form). However, the Shapley value is well-defined for non-superadditive games as well.

The Shapley value is based on the intuition that the payment that each agent receives should be *proportional to his contribution*. A naive implementation of this idea would be to pay each agent according to how much he increases the value of the coalition of all other players when he joins it, i.e., set the payoff of the player i to $v(N) - v(N \setminus \{i\})$. However, under this payoff scheme the total payoff assigned to the agents may differ from the value of the grand coalition: for instance, in the game described in Example 2.9 each agent's payoff under this scheme would be 0, whereas the value of the grand coalition is 1.

To avoid this problem, we can fix an ordering of the agents and pay each agent according to how much he contributes to the coalition formed by his predecessors in this ordering: that is, agent 1 receives $v(\{1\})$, agent 2 receives $v(\{1, 2\}) - v(\{1\})$, etc. It is easy to see that this payoff scheme distributes the value of the grand coalition among the agents. However, it suffers from another problem: two agents that play symmetric roles in the game may receive very different payoffs. Indeed, consider again the game described in Example 2.9: under the order-based payoff scheme one agent (the one that happens to be the second in the ordering) will receive 1, whereas two other agents receive 0, even though all three agents are clearly identical in terms of their contribution. In other words, the agents' payoffs in this scheme strongly depend on the selected ordering of the agents. Shapley's insight, which led to the definition of the Shapley value, was that this dependence can be eliminated by *averaging* over all possible orderings, or permutations, of the players.

To formally define the Shapley value, we need some additional notation. Fix a characteristic function game $G = (N, v)$. Let Π_N denote the set of all *permutations* of N, i.e., one-to-one mappings from N to itself. Given a permutation $\pi \in \Pi_N$, we denote by $S_\pi(i)$ the set of all predecessors of i in π, i.e., we set $S_\pi(i) = \{j \in N \mid \pi(j) < \pi(i)\}$. For example, if $N = \{1, 2, 3\}$ then

$$\Pi_N = \{(1, 2, 3), \ (1, 3, 2), \ (2, 1, 3), \ (2, 3, 1), \ (3, 1, 2), \ (3, 2, 1)\}.$$

Moreover, if $\pi = (3, 1, 2)$ then $S_\pi(3) = \emptyset$, $S_\pi(1) = \{3\}$, and $S_\pi(2) = \{1, 3\}$.

The *marginal contribution* of an agent i with respect to a permutation π in a game $G = (N, v)$ is denoted by $\Delta_\pi^G(i)$ and is given by

$$\Delta_\pi^G(i) = v(S_\pi(i) \cup \{i\}) - v(S_\pi(i)).$$

This quantity measures by how much i increases the value of the coalition consisting of its predecessors in π when he joins them. We can now define the Shapley value of a player i: it is simply his average marginal contribution, where the average is taken over all permutations of N.

Definition 2.10   Given a characteristic function game $G = (N, v)$ with $|N| = n$, the *Shapley value* of a player $i \in N$ is denoted by $\varphi_i(G)$ and is given by

$$\varphi_i(G) = \frac{1}{n!} \sum_{\pi \in \Pi_N} \Delta_\pi^G(i).$$

Example 2.11   Recall the ice cream game described in Example 2.2. In this game, the set of players is $N = \{C, M, P\}$, and for $c = 3$, $m = 4$, $p = 5$, the characteristic function v is given by $v(\emptyset) = 0$, $v(\{C\}) = v(\{M\}) = v(\{P\}) = 0$, $v(\{C, M\}) = v(\{C, P\}) = 500$, $v(\{M, P\}) = 750$, $v(\{C, M, P\}) = 1000$. Let us compute the Shapley value of player C (Charlie).

There are six permutations of the players: $\pi_1 = CMP, \pi_2 = CPM, \pi_3 = MCP, \pi_4 = PCM$, $\pi_5 = MPC$, and $\pi_6 = PMC$. We have

$$
\begin{aligned}
\Delta^G_{\pi_1}(C) &= v(\{C\}) - v(\emptyset) &&= 0 \\
\Delta^G_{\pi_2}(C) &= v(\{C\}) - v(\emptyset) &&= 0 \\
\Delta^G_{\pi_3}(C) &= v(\{C, M\}) - v(\{M\}) &&= 500 \\
\Delta^G_{\pi_4}(C) &= v(\{C, P\}) - v(\{P\}) &&= 500 \\
\Delta^G_{\pi_5}(C) &= v(N) - v(\{M, P\}) &&= 250 \\
\Delta^G_{\pi_6}(C) &= v(N) - v(\{M, P\}) &&= 250
\end{aligned}
$$

Hence, $\varphi_C(G) = (500 + 500 + 250 + 250)/6 = 250$.

The Shapley value has many attractive properties. First, it is *efficient*, i.e., it distributes the value of the grand coalition among all agents.

Proposition 2.12   *For any characteristic function game $G = (N, v)$ we have $\sum_{i=1}^{n} \varphi_i(G) = v(N)$.*

Proof. For any permutation π, the sum of the agents' marginal contributions with respect to π equals $v(N)$. Indeed, let $a_i = \pi^{-1}(i)$ for $i = 1, \ldots, n$: a_i is the player that appears in position i in π. Then we have

$$
\sum_{i=1}^{n} \Delta^G_\pi(i) = v(\{a_1\}) - v(\emptyset) + v(\{a_2, a_1\}) - v(\{a_1\})
$$
$$
+ \cdots + v(\{a_1, \ldots, a_n\}) - v(\{a_1, \ldots, a_{n-1}\}) = v(N).
$$

Thus, we obtain

$$
\sum_{i=1}^{n} \varphi_i(G) = \frac{1}{n!} \sum_{i=1}^{n} \sum_{\pi \in \Pi_N} \Delta^G_\pi(i) = \frac{1}{n!} \sum_{\pi \in \Pi_N} \sum_{i=1}^{n} \Delta^G_\pi(i) = \frac{1}{n!} \sum_{\pi \in \Pi_N} v(N) = v(N).
$$

$\square$

Another useful property of the Shapley value is that it does not allocate any payoffs to players who do not contribute to any coalition. Formally, given a characteristic function game $G = (N, v)$, a player $i \in N$ is said to be a *dummy* if $v(C) = v(C \cup \{i\})$ for any $C \subseteq N$. It is not hard to see that the Shapley value of a dummy player is 0.

Proposition 2.13   *Consider a characteristic function game $G = (N, v)$. If a player $i \in N$ is a dummy in G, then $\varphi_i(G) = 0$.*

Proof. Take an arbitrary permutation π. We have $v(S_\pi(i) \cup \{i\}) = v(S_\pi(i))$. Thus, $\Delta^G_\pi(i) = 0$. As this holds for any $\pi \in \Pi_N$, we have $\varphi_i(G) = 0$. $\square$

We remark that the converse is only guaranteed to be true if the game is monotone. Indeed, for a monotone game $G = (N, v)$, we have $\Delta_\pi^G(i) \geq 0$ for any $i \in N$ and any $\pi \in \Pi_N$. Thus, $\varphi_i(G) = \sum_{\pi \in \Pi_N} \Delta_\pi^G(i) > 0$ implies that $\Delta_\pi^G(i) > 0$ for some $\pi \in \Pi_N$, and hence $v(C \cup \{i\}) > v(C)$ for some $C \subseteq N$. On the other hand, consider a non-monotone game $G = (N, v)$ with $N = \{1, 2\}$, $v(\emptyset) = v(N) = 0$, $v(\{1\}) = v(\{2\}) = 2$. A direct calculation shows that $\varphi_1(G) = \varphi_2(G) = 0$ even though neither of the players is a dummy.

Another important observation is that if two players contribute equally to each coalition, then their Shapley values are equal. Formally, given a characteristic function game $G = (N, v)$, we say that players i and j are *symmetric* in G if $v(C \cup \{i\}) = v(C \cup \{j\})$ for any coalition $C \subseteq N \setminus \{i, j\}$. We will now show that symmetric players have equal Shapley values.

Proposition 2.14   *Consider a characteristic function game $G = (N, v)$. If players i and j are symmetric in G, then $\varphi_i(G) = \varphi_j(G)$.*

Proof.   Given a permutation π of N, let π' denote the permutation obtained by swapping i and j, i.e.,

$$\pi'(i) = \pi(j), \quad \pi'(j) = \pi(i), \quad \pi'(\ell) = \pi(\ell) \text{ for } \ell \neq i, j.$$

We claim that $\Delta_\pi^G(i) = \Delta_{\pi'}^G(j)$.

Indeed, suppose first that i precedes j in π. Then we have $S_\pi(i) = S_{\pi'}(j)$. Set $C = S_\pi(i) = S_{\pi'}(j)$. We obtain

$$\Delta_\pi^G(i) = v(C \cup \{i\}) - v(C), \quad \Delta_{\pi'}^G(j) = v(C \cup \{j\}) - v(C).$$

By symmetry, we have $v(C \cup \{i\}) = v(C \cup \{j\})$, which implies $\Delta_\pi^G(i) = \Delta_{\pi'}^G(j)$.

Now, suppose that i appears after j in π. Set $C = S_\pi(i) \setminus \{j\}$. We have

$$\Delta_\pi^G(i) = v(C \cup \{j\} \cup \{i\}) - v(C \cup \{j\}), \quad \Delta_{\pi'}^G(j) = v(C \cup \{j\} \cup \{i\}) - v(C \cup \{i\}).$$

Since C is a subset of N that contains neither i nor j, by symmetry we have $v(C \cup \{j\}) = v(C \cup \{i\})$ and therefore $\Delta_\pi^G(i) = \Delta_{\pi'}^G(j)$.

We have argued that $\Delta_\pi^G(i) = \Delta_{\pi'}^G(j)$ for any $\pi \in \Pi_N$. Now, observe that the mapping $\pi \to \pi'$ is one-to-one, and hence $\Pi_N = \{\pi' \mid \pi \in \Pi_N\}$. Therefore, we have

$$\varphi_i(G) = \frac{1}{n!} \sum_{\pi \in \Pi_N} \Delta_\pi^G(i) = \frac{1}{n!} \sum_{\pi \in \Pi_N} \Delta_{\pi'}^G(j) = \varphi_j(G).$$

$\square$

Finally, consider a group of players N that is involved in two coalitional games G^1 and G^2, i.e., $G^1 = (N, v^1)$, $G^2 = (N, v^2)$. The *sum* of G^1 and G^2 is a coalitional game $G^+ = G^1 + G^2$ given by $G^+ = (N, v^+)$, where for any coalition $C \subseteq N$ we have $v^+(C) = v^1(C) + v^2(C)$. It can easily be seen that the Shapley value of a player i in G^+ is the sum of his Shapley values in G^1 and G^2.

Proposition 2.15   *Consider two characteristic function games $G^1 = (N, v^1)$ and $G^2 = (N, v^2)$ over the same set of players N. Then for any player $i \in N$ we have $\varphi_i(G^1 + G^2) = \varphi_i(G^1) + \varphi_i(G^2)$.*

Proof. Let v^+ be the characteristic function of the game $G^1 + G^2$. Given a player $i \in N$ and a permutation π, let

$$\Delta_\pi^+(i) = v^+(S_\pi(i) \cup \{i\}) - v^+(S_\pi(i));$$

it is not hard to see that $\Delta_\pi^+(i) = \Delta_\pi^{G^1}(i) + \Delta_\pi^{G^2}(i)$. We obtain

$$\varphi_i(G^+) = \frac{1}{n!} \sum_{\pi \in \Pi_N} \Delta_\pi^+(i) = \sum_{\pi \in \Pi_N} (\Delta_\pi^{G^1}(i) + \Delta_\pi^{G^2}(i)) = \varphi_i(G^1) + \varphi_i(G^2),$$

and the proof is complete.   □

To summarize, we have shown that the Shapley value possesses four desirable properties:

(1) *Efficiency:* $\sum_{i \in N} \varphi_i(G) = v(N)$;

(2) *Dummy player:* if i is a dummy, then $\varphi_i(G) = 0$;

(3) *Symmetry:* if i and j are symmetric in G, then $\varphi_i(G) = \varphi_j(G)$;

(4) *Additivity:* $\varphi_i(G^1 + G^2) = \varphi_i(G^1) + \varphi_i(G^2)$ for all $i \in N$.

Now, it can be shown that the Shapley value is in fact the *only* payoff division scheme that has these four properties simultaneously. Thus, if we view properties (1)–(4) as axioms, these axioms *uniquely characterize* the Shapley value. To put it another way, if we want a payoff distribution scheme that satisfies these axioms, then the Shapley value will be such a scheme; and if we come up with a payoff distribution scheme that satisfies these axioms, then it is in fact nothing more than the Shapley value.

These properties can also be used to simplify the computation of the Shapley value.

Example 2.16   Consider a characteristic function game $G = (N, v)$ with $|N| = n$, where $v(N) = 1$ and $v(C) = 0$ for any $C \subset N$; this game is sometimes referred to as the *unanimity game*. We can compute the Shapley value of a player $i \in N$ by observing that $\Delta_\pi^G(i) = 1$ if i appears in the last position in π and $\Delta_\pi^G(i) = 0$ otherwise; since there are exactly $(n - 1)!$ permutations with player i in the last position, we have $\varphi_i(G) = \frac{(n-1)!}{n!} = \frac{1}{n}$. Alternatively, we can simply observe that all players in this game are symmetric, and hence $\varphi_1(G) = \cdots = \varphi_n(G)$; by the efficiency property, this implies $\varphi_i(G) = \frac{v(N)}{n} = \frac{1}{n}$ for all $i \in N$.

2.2.2 BANZHAF INDEX

Another solution concept that is motivated by fairness considerations is the Banzhaf index [40]. Just like the Shapley value, the Banzhaf index measures agents' expected marginal contributions; however, instead of averaging over all permutations of players, it averages over all coalitions in the game.

Definition 2.17 Given a characteristic function game $G = (N, v)$ with $|N| = n$, the *Banzhaf index* of a player $i \in N$ is denoted by $\beta_i(G)$ and is given by

$$\beta_i(G) = \frac{1}{2^{n-1}} \sum_{C \subseteq N \setminus \{i\}} [v(C \cup \{i\}) - v(C)].$$

It is not hard to verify that the Banzhaf index satisfies properties (2)–(4) in the list above. However, it does not satisfy property (1), i.e., efficiency, as the following example illustrates.

Example 2.18 Consider the unanimity game described in Example 2.16. Fix a player $i \in N$. We have $v(C \cup \{i\}) - v(C) = 1$ if $C = N \setminus \{i\}$ and $v(C \cup \{i\}) - v(C) = 0$ for any proper subset C of $N \setminus \{i\}$. Therefore, $\beta_i(G) = \frac{1}{2^{n-1}}$ for each $i \in N$ and hence $\sum_{i \in N} \beta_i(G) < v(N)$ for $n > 2$.

Since efficiency is surely a very desirable property of a payoff distribution scheme, a rescaled version of the Banzhaf index has been proposed. Formally, the *normalized Banzhaf index* $\eta_i(G)$ is defined as

$$\eta_i(G) = \frac{\beta_i(G)}{\sum_{i \in N} \beta_i(G)}.$$

While this version of the Banzhaf index satisfies efficiency, it loses the additivity property.

A few other power indices have been proposed in the literature; we refer the reader to [117] for an overview. However, these indices are not as popular as the Shapley value and the Banzhaf index.

2.2.2.1 Shapley Value and Banzhaf Index in Simple Games

In simple games, the Shapley value (which is often referred to as the *Shapley–Shubik power index* in this context) and the Banzhaf index have a particularly attractive interpretation: they measure the *power* of a player, i.e., the probability that she can influence the outcome of the game.

Indeed, the Shapley value of a player i in a simple game $G = (N, v)$ with $|N| = n$ can be rewritten as follows:

$$\varphi_i(G) = \frac{1}{n!} |\{\pi \in \Pi_N \mid v(S_\pi(i)) = 0, v(S_\pi(i) \cup \{i\}) = 1\}|.$$

To interpret this formula, we need an additional definition. In a simple game $G = (N, v)$, a player i is said to be *pivotal* for a coalition $C \subseteq N$ if $v(C) = 1, v(C \setminus \{i\}) = 0$. This definition can be extended to permutations: i is said to be pivotal for a permutation $\pi : N \to N$ if it is pivotal for the coalition $S_\pi(i) \cup \{i\}$ consisting of i itself and its predecessors in π. In this terminology, an agent i's Shapley value counts the fraction of permutations that i is pivotal for. In other words, if agents join the coalition in a random order, $\varphi_i(G)$ is exactly the probability that player i turns a losing coalition into a winning one. Similarly, the Banzhaf index measures the probability that a given agent turns a losing coalition into a winning one if each of the other agents decides whether to join the coalition by independently tossing a fair coin. It is this probabilistic interpretation that makes the original Banzhaf index a more attractive solution concept than its normalized cousin; this issue is discussed in detail by Felsenthal and Machover [118].

2.2.3   CORE AND CORE-RELATED CONCEPTS

We have introduced two solution concepts that attempt to measure the agents' marginal contribution. In contrast, the solution concepts considered in this and subsequent sections are defined in terms of coalitional stability.

Consider a characteristic function game $G = (N, v)$ and an outcome $(CS, \mathbf{x})$ of this game. Recall that $x(C)$ denotes the total payoff of a coalition C under $\mathbf{x}$. Now, if $x(C) < v(C)$ for some $C \subseteq N$, the agents in C could do better by abandoning the coalition structure CS and forming a coalition of their own. For example, they could distribute the additional payoff earned by this coalition by setting $x_i' = x_i + \frac{v(C) - x(C)}{|C|}$, i.e., share the extra profit equally among themselves: this ensures that each member of C has an incentive to deviate. Thus, in this case, the outcome $(CS, \mathbf{x})$ is unstable. The set of stable outcomes, i.e., outcomes where no subset of players has an incentive to deviate, is called the *core* of G [124].

Definition 2.19   The *core* $\mathcal{C}(G)$ of a characteristic function game $G = (N, v)$ is the set of all outcomes $(CS, \mathbf{x})$ such that $x(C) \geq v(C)$ for every $C \subseteq N$.

Example 2.20   Consider again the ice cream game of Example 2.2 with $c = 3$, $m = 4$, $p = 5$: we have $N = \{C, M, P\}$, $v(\emptyset) = 0$, $v(\{C\}) = v(\{M\}) = v(\{P\}) = 0$, $v(\{C, M\}) = v(\{C, P\}) = 500$, $v(\{M, P\}) = 750$, $v(\{C, M, P\}) = 1000$. Any outcome in the core of this game is of the form $(CS, (x_C, x_P, x_M))$. Moreover, since the value of the grand coalition is strictly greater than the social welfare of any other coalition structure, it has to be the case that $CS = \{N\}$: otherwise we would have $x(N) = \sum_{C \in CS} v(C) < v(N)$, violating the core constraint for N. Furthermore, the values x_C, x_P, x_M have to satisfy the following constraints:

$$x_C \geq 0, \quad x_M \geq 0, \quad x_P \geq 0$$
$$x_C + x_M + x_P = 1000$$
$$x_C + x_M \geq 500$$
$$x_C + x_P \geq 500$$
$$x_M + x_P \geq 750$$

This system has several solutions. For instance, $x_C = 0$, $x_M = x_P = 500$ produces an outcome in the core, and so does $x_C = 100$, $x_M = 400$, $x_P = 500$.

The outcomes in the core are stable in the sense that no coalition has any incentive to "defect"—no coalition can do better. Therefore, such outcomes are more likely to arise when a coalitional game is played.

The argument given in Example 2.20 shows that stability implies efficiency: any outcome in the core maximizes the *social welfare*, i.e., the total payoff of all players.

Proposition 2.21   *If an outcome* $(CS, \mathbf{x})$ *is in the core of a characteristic function game* $G = (N, v)$ *then* $v(CS) \geq v(CS')$ *for every coalition structure* $CS' \in \mathcal{CS}_N$.

Proof. Suppose for the sake of contradiction that $v(CS) < v(CS')$ for some coalition structure $CS' \in \mathcal{CS}_N$. Then we have

$$\sum_{C' \in CS'} x(C') = \sum_{i \in N} x_i = v(CS) < v(CS') = \sum_{C' \in CS'} v(C').$$

On the other hand, since $(CS, \mathbf{x})$ is in the core of G, we have $x(C') \geq v(C')$ for any $C' \in CS'$ and therefore $\sum_{C' \in CS'} x(C') \geq \sum_{C' \in CS'} v(C')$. This contradiction proves that CS maximizes the social welfare. □

Unfortunately, some games have empty cores.

Example 2.22   Let G be the 3-player majority game of Example 2.9. We claim that $\mathcal{C}(G) = \emptyset$. Indeed, suppose that the core of G is non-empty. Since $v(N) = 1$, any outcome $(CS, (x_1, x_2, x_3)) \in \mathcal{C}(G)$ satisfies $x_1 \geq 0$, $x_2 \geq 0$, $x_3 \geq 0$, and $x_1 + x_2 + x_3 \geq 1$. The latter constraint implies that $x_i \geq \frac{1}{3}$ for some $i \in \{1, 2, 3\}$.

On the other hand, we have $v(CS) \leq 1$ for any coalition structure $CS \in \mathcal{CS}_N$, and therefore $x_1 + x_2 + x_3 \leq 1$. Hence, for $C = N \setminus \{i\}$ we have $v(C) = 1$, $x(C) \leq 2/3$, which means that $(CS, (x_1, x_2, x_3))$ is not in the core. This contradiction shows that the core of G is empty.

2.2.3.1   The Core of Superadditive Games

Recall that in superadditive games we identify the set of outcomes with the set of payoff vectors for the grand coalition. Can this restriction eliminate some of the core outcomes, i.e., can it be the case that the core of a superadditive game contains an outcome where the grand coalition does not form?

Proposition 2.21 shows that if a game is *strictly superadditive*, i.e., if $v(N) > v(CS)$ for any coalition structure CS that consists of two or more coalitions, the answer is "no". On the other hand, if $v(N) = v(CS)$ for some $CS \in \mathcal{CS}_N$, the core may contain outcomes of the form $(CS, \mathbf{x})$. However, for any such outcome it holds that $(\{N\}, \mathbf{x})$ is in the core as well: we have

$$\sum_{i \in N} x_i = \sum_{C \in CS} \sum_{i \in C} x_i = \sum_{C \in CS} v(C) = v(N),$$

so $\mathbf{x}$ is a payoff vector for N, and all core constraints are trivially satisfied. This means that there is no loss of generality in ignoring core outcomes where the grand coalition does not form: for any such outcome there is an essentially equivalent outcome (i.e., with the same payoff vector) where the grand coalition forms. Therefore, for superadditive games we can simplify the notation and define the core as the set of all vectors $\mathbf{x}$ that satisfy:

(1) $x_i \geq 0$ for all $i \in N$;

(2) $x(N) = v(N)$; and

(3) $x(C) \geq v(C)$ for all $C \subseteq N$.

We remark, however, that if the game is not superadditive, it may be the case that the core of the game is non-empty, even though no outcome in which the grand coalition forms is stable. Thus, in non-superadditive games, requiring the grand coalition to form may cause a loss of stability.

Example 2.23 Consider a game $G = (N, v)$, where $N = \{1, 2, 3, 4\}$ and $v(C) = 0$ if $|C| < 2$, $v(C) = 2$ if $|C| \geq 2$. It is not hard to check that the outcome $(\{\{1, 2\}, \{3, 4\}\}, (1, 1, 1, 1))$ is in the core. However, no outcome of the form $(\{N\}, \mathbf{x})$ is in the core, since at least one of the coalitions $\{1, 2\}$ and $\{3, 4\}$ is paid less than 2 under $\mathbf{x}$ and therefore has an incentive to deviate.

Nevertheless, the following observation, whose proof is loosely based on [129], shows that there is no loss of generality in viewing the elements of the core as payoff vectors for the grand coalition: it suffices to replace the game with its superadditive cover.

Proposition 2.24 *A characteristic function game $G = (N, v)$ has a non-empty core if and only if its superadditive cover $G^* = (N, v^*)$ has a non-empty core.*

Proof. Suppose that $\mathcal{C}(G) \neq \emptyset$, and let $(CS, \mathbf{x})$ be an outcome in the core of G. By Proposition 2.21, we have $v^*(N) = v(CS)$, and therefore $\mathbf{x}$ is a payoff vector for the grand coalition in G^*. It remains to show that $\mathbf{x}$ satisfies all core constraints in G^*. Suppose for the sake of contradiction that $x(C) < v^*(C)$ for some coalition $C \subseteq N$. We have $v^*(C) = v(CS')$ for some coalition structure $CS' \in \mathcal{CS}_C$, which means that $x(C') < v(C')$ for some $C' \in CS'$, a contradiction with $(CS, \mathbf{x})$ being in the core of G. Hence, $\mathbf{x}$ is in $\mathcal{C}(G^*)$.

Conversely, suppose that $\mathcal{C}(G^*) \neq \emptyset$, and let $\mathbf{x}$ be an outcome in the core of G^*. Pick an arbitrary coalition structure CS such that $v(CS) = v^*(N)$. We have $x(N) = v(CS)$. On the other hand, since **x** satisfies the core constraints, we have $x(C) \geq v(C)$ for all $C \in CS$. By adding these inequalities together, we get $x(N) \geq v(CS)$, which means that each of these inequalities is, in fact, an equality, i.e., we have $x(C) = v(C)$ for all $C \in CS$. Hence, **x** is a payoff vector for CS. Moreover, since **x** is in the core of G^*, we have $x(C) \geq v^*(C) \geq v(C)$ for all $C \subseteq N$, and hence $(CS, \mathbf{x})$ is in $\mathcal{C}(G)$.   □

The argument in the proof of Proposition 2.24 does not necessarily extend to other solution concepts. For instance, paying the agents in G according to their Shapley values in the game G^* may require cross-coalitional transfers.

Example 2.25   Consider a 5-player game $G = (N, v)$, where $v(C) = 0$ if $|C| \in \{0, 1\}$, $v(C) = 6$ if $|C| = 2$, and $v(C) = 24$ if $|C| \geq 3$. We have $v^*(N) = 30$. Therefore, since all players are symmetric, the Shapley value of each player in the game G^* equals 5. However, in the game G, any optimal coalition structure contains a coalition of size 2 whose value is just 6.

We remark that much of the literature on coalitional games assumes superadditivity and justifies this assumption by saying that one can always replace the game with its superadditive cover. The example above illustrates that this approach may be problematic if cross-coalitional transfers are not allowed; see the influential paper of Aumann and Dreze [12] for a discussion of this issue. There is also another difficulty with this approach: computing the characteristic function of the game G^* may require a substantial computational effort; this issue is discussed at length in section 6.1.

2.2.3.2 The Core of Simple Games

Recall that for simple games it is standard to assume that the grand coalition forms, even if the game is not superadditive. Under this assumption, we can characterize the outcomes in the core, and provide a simple criterion for checking whether the game has a non-empty core. In what follows, we use the notation introduced in section 2.1.2.2 for superadditive games and identify an outcome of a game with a payoff vector $\mathbf{x} = (x_1, \ldots, x_n)$ for the grand coalitions.

We say that a player i in a simple game $G = (N, v)$ is a *veto* player if $v(C) = 0$ for any $C \subseteq N \setminus \{i\}$; since simple games are monotone, this is equivalent to requiring that $v(N \setminus \{i\}) = 0$. Observe that a game may have more than one veto player: for instance, in the unanimity game, where $v(N) = 1$, $v(C) = 0$ for any $C \subset N$, all players are veto players. We will now show that the only way to achieve stability in a simple game is to share the payoff among the veto players, if they exist.

Theorem 2.26   A simple game $G = (N, v)$ has a non-empty core if and only if it has a veto player. Moreover, an outcome $(x_1, \ldots, x_n)$ is in the core of G if and only if $x_i = 0$ for every player i who is not a veto player in G.

Proof. We will only prove the first statement of the theorem; the proof of the second statement is similar.

Suppose G has a veto player i. Then the outcome $\mathbf{x}$ with $x_i = 1$, $x_j = 0$ for $j \neq i$ is in the core: any coalition C that contains i satisfies $x(C) = 1 \geq v(C)$, whereas any coalition C' that does not contain i satisfies $v(C') = 0 \leq x(C')$.

Conversely, suppose that G does not have a veto player. Suppose for the sake of contradiction that G has a non-empty core, and let $\mathbf{x}$ be an outcome in the core of G. Since $x(N) = 1$, we have $x_i > 0$ for some $i \in N$, and hence $x(N \setminus \{i\}) = 1 - x_i < 1$. However, since i is not a veto player, we have $v(N \setminus \{i\}) = 1 > x(N \setminus \{i\})$, a contradiction with $\mathbf{x}$ being in the core. $\qquad\square$

The characterization of the outcomes in the core provided by Theorem 2.26 suggests a simple algorithm for checking if an outcome is in the core or deciding non-emptiness of the core: it suffices to determine, for each player i, whether he is a veto player, i.e., to compute $v(N \setminus \{i\})$. Thus, if the characteristic function of a simple game is efficiently computable, we can answer the core-related questions in polynomial time. It is important to note, however, that if we are given a simple game that is not superadditive, and we use the more general definition of a game outcome that we introduced at the beginning of this chapter (i.e., an outcome is a coalition structure together with a payoff vector), then Theorem 2.26 no longer holds. Moreover, deciding whether an outcome is in the core becomes computationally hard even for fairly simple representation formalisms (see chapter 4).

2.2.3.3 The Core of Convex Games

A classic result by Shapley [233] shows that convex games always have a non-empty core. Shapley's proof is constructive, i.e., it shows how to obtain an outcome in the core of a convex game.

Theorem 2.27   If $G = (N, v)$ is a convex game, then G has a non-empty core.

Proof. Fix an arbitrary permutation $\pi \in \Pi_N$, and let x_i be the marginal contribution of i with respect to π, i.e., set $x_i = \Delta_\pi^G(i)$. We claim that $(x_1, \ldots, x_n)$ is in the core of G.

Indeed, observe first that any convex game is monotone, so $x_i \geq 0$ for all $i \in N$. Moreover, a telescoping argument similar to the one used in the proof of Proposition 2.12 shows that $\sum_{i \in N} x_i = v(N)$. Now, suppose for the sake of contradiction that we have $v(C) > x(C)$ for some coalition $C = \{i_1, \ldots, i_s\}$. We can assume without loss of generality that $\pi(i_1) \leq \cdots \leq \pi(i_s)$, i.e., the members of C appear in π ordered as $i_1, \ldots, i_s$. We can write $v(C)$ as

$$v(C) = v(\{i_1\}) - v(\emptyset) + v(\{i_1, i_2\}) - v(\{i_1\}) + \cdots + v(C) - v(C \setminus \{i_s\}).$$

Now, for each $j = 1, \ldots, s$ we have $\{i_1, \ldots, i_{j-1}\} \subseteq \{1, \ldots, i_j - 1\}$, so the supermodularity of v implies

$$v(\{i_1, \ldots, i_{j-1}, i_j\}) - v(\{i_1, \ldots, i_{j-1}\}) \leq v(\{1, \ldots, i_j\}) - v(\{1, \ldots, i_j - 1\}) = x_{i_j}.$$

By adding up these inequalities for $j = 1, \ldots, s$, we obtain $v(C) \leq x(C)$, i.e., coalition C does not have an incentive to deviate, a contradiction. $\qquad\square$

Observe that the construction used in the proof of Theorem 2.27 immediately implies that in a convex game the Shapley value is in the core: indeed, the Shapley value is a convex combination of outcomes constructed in the proof of Theorem 2.27, and the core can be shown to be a convex set. However, Theorem 2.27 does not, in general, enable us to check whether a given outcome is in the core of a given convex game.

2.2.3.4   Least Core

The notion of stability captured in the core requires that no coalition can gain *anything* by deviating. Conversely, an outcome is unstable if a coalition can benefit even by a tiny amount from deviating. This is a fairly strong requirement. In many situations, a deviation may be costly, and it would only make sense for a coalition to deviate if the gain from a deviation exceeded the costs of deviation. So, we might naturally consider relaxing the notion of the core, and only require that no coalition can benefit *significantly* by deviating. This motivates the following definition[1]:

Definition 2.28   An outcome $\mathbf{x}$ is said to be in the ε-*core* of a superadditive game G for some $\varepsilon \in \mathbb{R}$ if $x(C) \geq v(C) - \varepsilon$ for each $C \subseteq N$.

Clearly, the core of a game is exactly its ε-core for $\varepsilon = 0$. In practice we would usually be interested in finding the smallest value of ε such that the ε-core is non-empty. The corresponding ε-core is called the *least core* of G [176]. More formally, we have the following definition.

Definition 2.29   Given a superadditive game G, let

$$\varepsilon^*(G) = \inf\{\varepsilon \mid \varepsilon\text{-core of } G \text{ is non-empty}\}.$$

The *least core* of G is its $\varepsilon^*(G)$-core. The quantity $\varepsilon^*(G)$ is called the *value of the least core* of G.

It is not hard to see that the set $\{\varepsilon \mid \varepsilon\text{-core of } G \text{ is non-empty}\}$ is closed and therefore contains its greatest lower bound, i.e., the least core is always non-empty; an alternative proof of this fact will be provided in section 2.2.3.6. This means that we can replace inf in the definition of the least core with min.

Example 2.30   Let G be the 3-player majority game of Example 2.9. In Example 2.22 we have argued that the core of this game is empty. In fact, the argument given in Example 2.22 shows that $\varepsilon^*(G) \geq \frac{1}{3}$, as for any payoff vector there exists a player that is paid at least $\frac{1}{3}$ and hence a winning coalition that is paid at most $\frac{2}{3}$. On the other hand, it is immediate that $(\frac{1}{3}, \frac{1}{3}, \frac{1}{3})$ is in the $\frac{1}{3}$-core of G. Thus, the value of the least core of G is $1/3$.

Observe that if G has a non-empty core, it may happen that $\varepsilon^*(G) < 0$, in which case the least core is a subset of the core. We remark, however, that some authors require the value of the

---

[1]Here (and for the rest of this section) we limit ourselves to superadditive games, but all of the definitions that follow can be extended to the general case.

least core to be non-negative, i.e., they define the least core as the smallest *non-negative* value of ε for which the ε-core is non-empty.

2.2.3.5 Cost of Stability

When defining the least core, we relaxed the stability constraints in the definition of the core. An alternative approach is to relax the feasibility constraint. This corresponds to having a benevolent external party that wishes to *stabilize* the game, by offering subsidies to players if they stay in the grand coalition. Following Bachrach *et al.* [27], we formalize this approach as follows.

Definition 2.31   Given a superadditive game $G = (N, v)$ and a $\Delta \geq 0$, let $G^\Delta = (N, v^\Delta)$ be a characteristic function game over the set of players N given by $v^\Delta(N) = v(N) + \Delta, v^\Delta(C) = v(C)$ for all $C \subset N$. The *cost of stability (CoS)* of G is the smallest (nonnegative) value of Δ such that G^Δ has a non-empty core; we will denote it by $\text{CoS}(G)$.

Example 2.32   Let G be the 3-player majority game of Example 2.9. We claim that $\text{CoS}(G) = \frac{1}{2}$. Indeed, for $\Delta = \frac{1}{2}$ the outcome $(\frac{1}{2}, \frac{1}{2}, \frac{1}{2})$ is in the core of G^Δ. On the other hand, if $\Delta < \frac{1}{2}$, in any outcome of G^Δ there exists a player who is paid at least $\frac{1+\Delta}{3}$, and hence there exists a winning coalition that is paid at most $\frac{2(1+\Delta)}{3} < 1$.

Example 2.33   Consider a variant of the ice cream game where $c = m = p = 5.5$. Then for any coalition $S \subseteq N$ we have $v(S) = 0$ if $|S| < 2, v(S) = 1000$ if $|S| = 2, v(S) = 1250$ if $|S| = 3$. It is not hard to check that this game has an empty core, i.e., the children cannot agree on a fair division of the ice cream. However, to avoid fighting, Charlie's parents can offer the children additional $\Delta = 250$g of ice cream from their freezer if the children agree to cooperate. The resulting game G^Δ has a non-empty core: in particular, $x_C = x_M = x_P = 500$ is in the core. Thus, the cost of stability of this game is at most 250; in fact, it is not hard to verify that it is exactly 250.

This idea can be extended to non-superadditive games. However, in this case it is less clear what the goal of the benevolent external party should be: while in some settings it may still want to prop up the grand coalition, in other cases it may look for a coalition structure that would be the cheapest to stabilize by external payments.

It is easy to see that the cost of stability can be bounded as:

$$0 \leq \text{CoS}(G) \leq n \max_{C \subseteq N} v(C).$$

Observe also that the cost of stability is related to the value of the least core: we have $\varepsilon^*(G) > 0$ if and only if $\text{CoS}(G) > 0$. However, there are examples where the cost of stability exceeds the value of the least core by a factor of n. To get a better understanding of how these two notions are related, we will now characterize both of them in terms of linear programming.

2.2.3.6 Core, Least Core and Cost of Stability: a Linear Programming Perspective

Note that this section assumes some understanding of linear programming (see, e.g., [204, 226]).

The set of all outcomes in the core of a superadditive game can be characterized by the following *linear feasibility program* (LFP).

$$
\begin{aligned}
x_i &\geq 0 \quad \text{for each } i \in N \\
\sum_{i \in N} x_i &= v(N) \\
\sum_{i \in C} x_i &\geq v(C) \quad \text{for each } C \subseteq N
\end{aligned}
\tag{2.1}
$$

This LFP has $2^n + n + 1$ constraints. If we want to convert it into an algorithm for checking non-emptiness of the core that runs in time polynomial in n, then we need an efficient *separation oracle* for this LFP. A separation oracle for a linear (feasibility) program is a procedure that, given a candidate solution $(x_1, \ldots, x_n)$ determines whether it is feasible and, if not, outputs the violated constraint. It is well-known that if a linear program over n variables admits a separation oracle that runs in time $\text{poly}(n)$, then an optimal feasible solution can be found it time $\text{poly}(n)$ [226].

Now, the first $n + 1$ constraints in our LFP are straightforward to check. Therefore, the problem of checking non-emptiness of the core for superadditive games can be reduced to checking whether a candidate solution satisfies the last 2^n constraints, i.e., verifying whether a given outcome is in the core (and, if not, computing the coalition that has an incentive to deviate).

By modifying the LFP (2.1) slightly, we can obtain a linear program for computing the value of the least core as well as an outcome in the least core; this provides an alternative proof that the least core is always non-empty. Indeed, it is not hard to see that the optimal solution to the following linear program with variables $\varepsilon, x_1, \ldots, x_n$ is exactly the value of the least core:

$$
\begin{aligned}
\min \quad & \varepsilon \\
& x_i \geq 0 \quad \text{for each } i \in N \\
& \sum_{i \in N} x_i = v(N) \\
& \sum_{i \in C} x_i \geq v(C) - \varepsilon \quad \text{for each } C \subseteq N
\end{aligned}
\tag{2.2}
$$

Similarly, we can design a linear program to compute the cost of stability. We have:

$$
\begin{aligned}
\min \quad & \Delta \\
x_i \; &\geq \; 0 \quad \text{for each } i \in N \\
\sum_{i \in N} x_i \; &= \; v(N) + \Delta \\
\sum_{i \in C} x_i \; &\geq \; v(C) \quad \text{for each } C \subseteq N
\end{aligned}
\tag{2.3}
$$

In many cases, a separation oracle for the LFP (2.1) can be modified to work for (2.2) and (2.3); we will see examples of this in chapter 4.

2.2.3.7 The Bondareva–Shapley Theorem

By replacing the objective function of LP (2.3) with $\sum_{i \in N} x_i$ and taking the dual of the resulting LP, we can derive a complete characterization of superadditive games that have a non-empty core. This observation was first made by Bondareva [57] and, independently, by Shapley [232], and the resulting characterization became known as the Bondareva–Shapley theorem. To state this theorem, we need a few additional definitions.

Given a set $S \subseteq N$, let $\chi_S : N \to \{0, 1\}$ be the *indicator function* of the set S, i.e., let $\chi_S(i) = 1$ if $i \in S$ and $\chi_S(i) = 0$ if $i \in N \setminus S$. A collection of sets $\mathcal{S} \subseteq 2^N \setminus \{\emptyset\}$ is said to be *balanced* if there exists a vector $(\delta_S)_{S \in \mathcal{S}}$ of positive numbers such that

$$
\sum_{S \in \mathcal{S}} \delta_S \chi_S = \chi_N.
$$

The vector $(\delta_S)_{S \in \mathcal{S}}$ is called the *balancing weight system* for $\mathcal{S}$. Observe that any coalition structure CS over N is a balanced set system with $\delta_S = 1$ for any $S \in CS$. Indeed, balanced set systems can be viewed as generalized partitions of N. We are now ready to state the theorem.

Theorem 2.34 (Bondareva–Shapley)   A superadditive game $G = (N, v)$ has a non-empty core if and only if for every balanced collection of sets $\mathcal{S} \subseteq 2^N \setminus \{\emptyset\}$ and any balancing weight system $(\delta_S)_{S \in \mathcal{S}}$ for $\mathcal{S}$ it holds that

$$
\sum_{S \in \mathcal{S}} \delta_S v(S) \leq v(N).
$$

We remark that, in general, the Bondareva–Shapley theorem does not provide an efficient method of checking whether a given game has a non-empty core, as the number of balanced set systems is superexponential in the number of agents n.

2.2.4   NUCLEOLUS

The *nucleolus* [225] is a solution concept that defines a unique outcome for a game. It can be thought of as a refinement of the least core. The nucleolus is based on the notion of *deficit*. Formally, given a superadditive game $G = (N, v)$, a coalition $C \subseteq N$, and a payoff vector $\mathbf{x}$ for this game, the *deficit* of C with respect to $\mathbf{x}$ is defined as $d(\mathbf{x}, C) = v(C) - x(C)$: this quantity measures C's incentive to deviate under $\mathbf{x}$. Any payoff vector $\mathbf{x}$ generates a 2^n-dimensional *deficit vector* $\mathbf{d}(\mathbf{x}) = (d(\mathbf{x}, C_1), \dots, d(\mathbf{x}, C_{2^n}))$, where $C_1, \dots, C_{2^n}$ is the list of all subsets of N ordered by their deficit under $\mathbf{x}$, from the largest to the smallest: $v(C_i) - x(C_i) \geq v(C_j) - x(C_j)$ for any $1 \leq i < j \leq 2^n$. Two deficit vectors can be compared lexicographically: given two payoff vectors $\mathbf{x}, \mathbf{y}$, we say that $\mathbf{d}(\mathbf{x})$ is *lexicographically smaller* than $\mathbf{d}(\mathbf{y})$ if there exists an $i \in \{1, \dots, 2^n\}$ such that the first $i - 1$ entries of $\mathbf{d}(\mathbf{x})$ and $\mathbf{d}(\mathbf{y})$ are equal, but the i-th entry of $\mathbf{d}(\mathbf{x})$ is smaller than the i-th entry of $\mathbf{d}(\mathbf{y})$; if this is the case, we write $\mathbf{d}(\mathbf{x}) <_{\text{lex}} \mathbf{d}(\mathbf{y})$. We extend this notation by setting $\mathbf{d}(\mathbf{x}) \leq_{\text{lex}} \mathbf{d}(\mathbf{y})$ if $\mathbf{d}(\mathbf{x}) <_{\text{lex}} \mathbf{d}(\mathbf{y})$ or $\mathbf{d}(\mathbf{x}) = \mathbf{d}(\mathbf{y})$. With this notion at hand, we can define the nucleolus: it is the set of all imputations that have the lexicographically smallest deficit vector.

Definition 2.35   The *nucleolus* $\mathcal{N}(G)$ of a superadditive game is the set

$$\mathcal{N}(G) = \{\mathbf{x} \in \mathcal{I}(N) \mid \mathbf{d}(\mathbf{x}) \leq_{\text{lex}} \mathbf{d}(\mathbf{y}) \text{ for all } \mathbf{y} \in \mathcal{I}(N)\}.$$

If, instead of optimizing over all imputations, we optimize over all payoff vectors, we obtain the *pre-nucleolus*; this distinction becomes immaterial if we assume that the value of any singleton coalition is 0.

Definition 2.35 defines the nucleolus as a set, and does not rule out the possibility that this set is empty. However, it can be shown that this is never the case: for any superadditive game, the set $\mathcal{N}(G)$ is non-empty, and, moreover, contains exactly one element. This remarkable property makes the nucleolus a very attractive solution concept, as it specifies a unique way to divide the payoff of the grand coalition.

Now, assume that $v(C) = 0$ if $|C| \leq 1$, i.e., the nucleolus coincides with the pre-nucleolus. Under this constraint, the nucleolus is guaranteed to be in the least core. To see this, observe that the least core can be defined as the set of all payoff vectors $\mathbf{x}$ that minimize the maximum deficit $d_1 = \max\{d(\mathbf{x}, C) \mid C \subseteq N\}$. By definition, this set contains the (pre)-nucleolus: indeed, the (pre)-nucleolus minimizes the largest deficit, then minimizes the second-largest deficit, etc.

The direct procedure for computing the (pre-)nucleolus involves computing the exponentially long deficit vector, and is therefore infeasible for all but the smallest coalitional games. However, the previous paragraph suggests a different approach [176]. Namely, we start by computing the set of payoff vectors that minimize the largest deficit d_1, i.e., the least core (using, e.g., the LP (2.2)). Then, among all payoff vectors in the least core, we pick the ones that minimize the second largest deficit $d_2 = \max\{d(\mathbf{x}, C) \mid C \subseteq N, d(\mathbf{x}, C) < d_1\}$, and remove all other payoff vectors. We can continue this procedure until the surviving set stabilizes. It is easy to see that at this point we obtain the pre-nucleolus (and, if all singleton coalitions have value 0, the nucleolus). Kopelowitz [157] shows that a variant of this procedure converges in at most n steps. However, at each step we need to solve an exponentially large linear program, so transforming this high-level description into an efficient procedure for computing the nucleolus is still a highly non-trivial task. Nevertheless, some classes of games defined on combinatorial structures (see chapters 3 and 4) admit efficient algorithms for computing the nucleolus: see, e.g. [95, 109, 150].

2.2.5   KERNEL

The *kernel* [86] consists of all outcomes where no player can credibly demand a fraction of another player's payoff. Formally, for any player i we define his *surplus* over the player j with respect to a payoff vector $\mathbf{x}$ as the quantity

$$S_{i,j}(\mathbf{x}) = \max\{v(C) - x(C) \mid C \subseteq N, i \in C, j \notin C\}.$$

Intuitively, this is the amount that player i can earn without the cooperation of player j, by asking a set $C \setminus \{i\}$ to join him in a deviation, and paying each player in $C \setminus \{i\}$ what it used to be paid under $\mathbf{x}$. Now, if $S_{i,j}(\mathbf{x}) > S_{j,i}(\mathbf{x})$, player i should be able to demand a fraction of player j's payoff—unless player j already receives the smallest payment that satisfies the individual rationality condition, i.e., $v(\{j\})$. Following this intuition, we define the *kernel* of a superadditive game G as follows:

Definition 2.36   The kernel $\mathcal{K}(G)$ of a superadditive game G is the set of all imputations $\mathbf{x}$ such that for any pair of players (i, j) we have either:

(1) $S_{i,j}(\mathbf{x}) = S_{j,i}(\mathbf{x})$; or

(2) $S_{i,j}(\mathbf{x}) > S_{j,i}(\mathbf{x})$ and $x_j = v(\{j\})$; or

(3) $S_{i,j}(\mathbf{x}) < S_{j,i}(\mathbf{x})$ and $x_i = v(\{i\})$.

Schmeidler [225] shows that the kernel always contains the nucleolus, and therefore it is guaranteed to be non-empty.

2.2.6   BARGAINING SET

The *bargaining set* [13] is defined similarly to the core. However, in contrast to the core, it only takes into account coalitional deviations that are themselves stable, i.e., do not admit a counterdeviation. There are several slightly different definitions of this concept; we present the one proposed by Mas-Colell [173].

Consider a superadditive game $G = (N, v)$ with $|N| = n$ and a payoff vector $\mathbf{x}$ for this game. A pair $(\mathbf{y}, C)$ is said to be an *objection* to $\mathbf{x}$ if C is a subset of N and $\mathbf{y}$ is a vector in $\mathbb{R}^n$ that satisfies $y(C) \le v(C)$, $y_i \ge x_i$ for each $i \in C$, and $y_i > x_i$ for at least one $i \in C$. A pair $(\mathbf{z}, D)$ is said to be a *counterobjection* to an objection $(\mathbf{y}, C)$ if D is a subset of N and $\mathbf{z}$ is a vector in $\mathbb{R}^n$ that satisfies:

(1) $z(D) \leq v(D)$;

(2) $z_i \geq y_i$ for all $i \in D \cap C$; and

(3) $z_i \geq x_i$ for all $i \in D \setminus C$,

with at least one of the inequalities in (2) and (3) being strict. An objection is said to be *justified* if it does not admit a counterobjection. Now, we are ready to define the bargaining set.

Definition 2.37　The bargaining set $\mathcal{B}(G)$ of a superadditive game $G = (N, v)$ is the set of all payoff vectors that do not admit a justified objection.

It is not hard to see that the core of a superadditive game is exactly the set of all payoff vectors that do not admit an objection (justified or otherwise). Consequently, the bargaining set contains the core, and the containment is sometimes strict. In fact, Einy *et al.* [104] show that the bargaining set contains the least core, which implies that the bargaining set is guaranteed to be non-empty. The bargaining set also contains the kernel [86].

2.2.7　STABLE SET

The *stable set* is the very first solution concept that was proposed for characteristic function games: it was put forward by von Neumann and Morgenstern in their seminal book [257], and is sometimes referred to as the *von Neumann–Morgenstern solution*. Similarly to the core or the kernel, it is a set-valued solution concept. However, a game may have several stable sets[2].

The definition of a stable set is based on the idea of *dominance*. Given a superadditive game $G = (N, v)$ and two imputations $\mathbf{y}, \mathbf{z} \in \mathcal{I}(N)$, we say that $\mathbf{y}$ *dominates* $\mathbf{z}$ via a coalition $C \subseteq N$ if $y(C) \leq v(C)$ and $y_i > z_i$ for all $i \in C$; we write $\mathbf{y} \, \text{dom}_C \, \mathbf{z}$ whenever this is the case. Note that if $\mathbf{y}$ dominates $\mathbf{z}$ via C, then $(\mathbf{y}, C)$ is an objection to $\mathbf{z}$, but the converse is not necessarily true. If $\mathbf{y}$ dominates $\mathbf{z}$ via some non-empty coalition $C \subseteq N$, we say that $\mathbf{y}$ *dominates* $\mathbf{z}$ and write $\mathbf{y} \, \text{dom} \, \mathbf{z}$. In other words, $\mathbf{y} \, \text{dom} \, \mathbf{z}$ if and only if there exists a $C \in 2^N \setminus \{\emptyset\}$ such that $\mathbf{y} \, \text{dom}_C \, \mathbf{z}$. Further, given a set of imputations $J \subseteq \mathcal{I}(N)$, let

$$\text{Dom}(J) = \{\mathbf{z} \in \mathcal{I}(N) \mid \text{there exists a } \mathbf{y} \in J \text{ such that } \mathbf{y} \, \text{dom} \, \mathbf{z}\}.$$

That is, if an imputation $\mathbf{z}$ is in $\text{Dom}(J)$, then there exists a non-empty set of players $C \subseteq N$ and an imputation $\mathbf{y} \in J$ such that each player in C prefers $\mathbf{y}$ to $\mathbf{z}$. We are now ready to present the definition of a stable set.

Definition 2.38　Given a superadditive game $G = (N, v)$, a set of imputations J is called a *stable set* of G if $\{J, \text{Dom}(J)\}$ is a partition of $\mathcal{I}(N)$, i.e.,

---

[2]This means that the stable set does not quite fit the definition of solution concept given in chapter 1, in that the mapping from games to the set of all outcomes that it defines is multi-valued.

- $J \cap \mathrm{Dom}(J) = \emptyset$ (internal stability), and

- $\mathcal{I}(N) \setminus J \subseteq \mathrm{Dom}(J)$ (external stability).

Even though a game can have many stable sets, there are also games that have no stable sets, though constructing such games is not trivial; the first example of a game with no stable set is due to Lucas [170].

It is not hard to see that if the core of a superadditive game G is non-empty, then it is contained in all stable sets of G: no imputation in the core is dominated by any other imputation. If G is convex, it has a unique stable set, which coincides with the core [233].

CHAPTER 3

Representations and Algorithms

When defining characteristic function games, we observed that the naive representation for such games, which lists every coalition together with its value, is exponential in the number of players, and is therefore impractical for most real-life scenarios. Thus, a more succinct representation formalism is needed. However, a simple counting argument shows that no language can be universally succinct, i.e., encode every n-player characteristic function game using $\text{poly}(n)$ bits. There are two ways of resolving this apparent contradiction:

(1) We can focus on subclasses of games that are defined via "small" combinatorial structures. While such a representation may fail to be universally expressive (i.e., there may exist characteristic function games that cannot be represented in this manner), it is nevertheless guaranteed to be succinct. This approach has received much attention in the theoretical computer science and operations research literature; we will discuss it in more detail in section 3.1.

(2) We can develop universally expressive representation languages that are succinct for *some* useful subclasses of coalitional games, and characterize the games that are likely to be compactly representable in these languages. Several languages in this category have been recently proposed in the multiagent systems literature; we survey this line of work in section 3.2.

These two strategies need not be mutually exclusive: this point is illustrated by chapter 4, where we study in detail a specific representation formalism that is succinct, but not universal (weighted voting games), and then show that a generalization of this formalism (vector weighted voting games) is universally expressive for the class of all simple games.

A closely related question is that of efficiently computing solution concepts for a given game. The running time of an algorithm is usually measured as a function of its input size, so the performance of a given procedure for computing a solution concept crucially depends on the underlying representation formalism. For instance, a trivial algorithm for checking if a given outcome is in the core, which looks at every coalition to check whether it has an incentive to deviate, runs in polynomial time with respect to the naive representation, but its running time will be exponential if the input is represented using $\text{poly}(n)$ bits (where n is the number of players). Therefore, for each representation language one needs to develop a set of algorithms for efficiently computing stable and/or fair outcomes, or prove that such algorithms are unlikely to exist. In this chapter, we present such algorithmic results for the models we consider. We remark that, unless explicitly indicated otherwise, throughout this section we identify the core of the game with the set of stable payoff vectors for the grand coalition, i.e., we do not consider games with coalition structures.

More ambitiously, we may ask what can be said about the complexity of computing a solution concept if all we know about the underlying representation is that it admits an efficient algorithm that given a coalition, outputs its value; this model is sometimes referred to as the *black-box* model, or the *oracle* model. In section 3.3, we give a brief overview of known complexity results in this very general model.

3.1   COMBINATORIAL OPTIMIZATION GAMES

In this section, we discuss several classes of games that belong to the category of *combinatorial optimization games*. These are games that are defined via a combinatorial structure (typically, a graph). The value of each coalition is obtained by solving a combinatorial optimization problem on the substructure that corresponds to this coalition. An excellent (though by now somewhat outdated) survey of combinatorial optimization games can be found in [59].

3.1.1   INDUCED SUBGRAPH GAMES

The first representation we look at was introduced by Deng and Papadimitriou [97]. The idea of the representation is very simple. A game is described by an undirected, weighted graph $\mathcal{G} = (N, E)$ with $|N| = n$ and $|E| = m$. The weight of edge (i, j) is denoted by $w_{i,j}$; we write $\mathbf{w} = (w_{i,j})_{i,j \in N}$. Without loss of generality we assume that all weights are integer (though not necessarily positive). In the game $G(\mathcal{G}, \mathbf{w})$ that corresponds to this graph, the set of players is N, and the value of a coalition $C \subseteq N$ is defined to be the total weight of all its internal edges, or, in other words, the *subgraph induced by* C:

$$v(C) = \sum_{\substack{(i, j) \in E \\ \{i, j\} \subseteq C}} w_{i,j}.$$

We allow self-loops, so that the value of a singleton coalition may be any integer number.

Figure 3.1 illustrates the idea. In Figure 3.1(a), we see the basic graph defining a characteristic function for four agents, $\{A, B, C, D\}$. Figure 3.1(b) shows the subgraph induced by the coalition $\{A, C, D\}$. From this, we can see that $v(\{A, B, C, D\}) = 10$, while $v(\{A, C, D\}) = 2 + 4 = 6$.

Induced subgraph games can be used to model social networks, where the value of each coalition (team, club) is determined by the relationships among its members: a player assigns a positive utility to being in a coalition with his friends and a negative utility to being in a coalition with his enemies.

Clearly, the induced subgraph representation is succinct as long as the number of bits required to encode edge weights is polynomial in $n = |N|$: using an adjacency matrix to represent the graph requires only n^2 entries. Of course, this is not a *complete* representation: in any induced subgraph

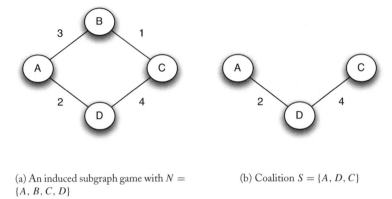

(a) An induced subgraph game with $N = \{A, B, C, D\}$

(b) Coalition $S = \{A, D, C\}$

Figure 3.1: Induced subgraph game representation.

game, the value of any coalition is completely determined by the values of its subcoalitions of size one and two, and in general this is not the case. This is illustrated by the following example.

Example 3.1 Consider a three-player game $G = (N, v)$, where $N = \{1, 2, 3\}$ and

$$v(C) = \begin{cases} 0 & \text{if } |C| \leq 1; \\ 1 & \text{if } |C| = 2; \\ 6 & \text{if } |C| = 3. \end{cases}$$

We claim that this game does not have an induced subgraph representation. Indeed, suppose that there exists a graph $\mathcal{G}$ and a weight vector $\mathbf{w}$ such that $G = G(\mathcal{G}, \mathbf{w})$. In $\mathcal{G}$, the weight of each self-loop would have to be 0, and every other edge would have to be present in $\mathcal{G}$ and have weight 1; but then the value of the grand coalition would be 3, and not 6.

Since we allow for negative edge weights, induced subgraph games are not necessarily monotone. However, when all edge weights are non-negative, induced subgraph games are not only monotone, but convex (this is easily verified using Proposition 2.8), and hence are guaranteed to have a non-empty core. Moreover, Deng and Papadimitriou describe an efficient network flow-based algorithm for checking whether an outcome is in the core of an induced subgraph game. In contrast, if weights may be negative, deciding whether the core is empty is NP-complete, whereas checking whether a specific outcome is in the core is coNP-complete [97]. Intuitively, these problems are closely related to the classic MAXCUT problem, which is well-known to be computationally hard. Furthermore, Greco *et al.* [128] have recently shown that checking if an outcome is in the kernel is Δ_2^p-complete, whereas checking if an outcome is in the bargaining set is Π_2^p-complete.

Given these hardness results, one might expect that computing the Shapley value in induced subgraph games would be difficult as well. However, this intuition turns out to be incorrect: Deng and Papadimitriou show that induced subgraph games admit an efficient algorithm for Shapley value calculation, which makes use of the axiomatic characterization of the Shapley value given in chapter 2. We will now present this algorithm, together with a proof of correctness; similar ideas can be used to compute the Shapley value in several other classes of coalitional games (most notably, MC-nets, see section 3.2.1).

Consider an induced subgraph game G given by a graph $\mathcal{G} = (N, E)$ and a weight vector $\mathbf{w}$, and let $\{e_1, \ldots, e_m\}$ be the set of edges of this graph. We can decompose the graph $\mathcal{G}$ into m graphs $\mathcal{G}^1, \ldots, \mathcal{G}^m$, where for every $j = 1, \ldots, m$ the graph $\mathcal{G}^j$ has the vertex set N and a single edge e_j (with the same weight as in the original graph); let G^j denote the game that corresponds to the j-th graph in this sequence. The reader can easily convince herself that

$$G(\mathcal{G}, \mathbf{w}) = G^1 + \cdots + G^m.$$

Therefore, by the additivity axiom for each player $i \in N$ we have $\varphi_i(G) = \sum_{j=1}^{m} \varphi_i(G^j)$, and it remains to compute the Shapley value of player i in the game G^j.

Suppose first that i is not incident to e_j. Then i is clearly a dummy in G^j and therefore we have $\varphi_i(G^j) = 0$. On the other hand, suppose that $e_j = (i, \ell)$ for some $\ell \in N$. Then i and ℓ are symmetric in G^j. Since the value of the grand coalition in G^j equals $w_{i,\ell}$, by efficiency and symmetry we get $\varphi_i(G^j) = w_{i,\ell}/2$.

Putting these observations together, we conclude that an agent's Shapley value is *half the income from the edges in the graph to which it is attached*, that is:

$$\varphi_i(G) = \frac{1}{2} \sum_{(i,j) \in E} w_{i,j}.$$

We remark that Deng and Papadimitriou show that in induced subgraph games the (pre)nucleolus coincides with the Shapley value; thus, the expression above can also be used to compute the (pre)nucleolus.

3.1.2   NETWORK FLOW GAMES

In *network flow games* [148, 149], the players are edges of a network with a source s and a sink t. Each edge e has a positive integer capacity c_e, indicating how much flow it can carry. The value of a coalition C is the maximum amount of flow that can be sent from s to t using the edges in C only. These games can model computer networks, in which case nodes in the graph correspond to computers or network routers, edges correspond to network connections, and capacities to the number of bits per second that can be transmitted along the connection. The agents that control the edges might be the enterprises that are responsible for maintaining the corresponding network links. Alternatively, network flow games can model a road network, in which case nodes correspond to intersections, and the capacity of a link shows the number of vehicles that can travel through this

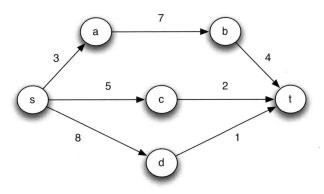

Figure 3.2: A network flow game.

link in a given amount of time (e.g., an hour). Figure 3.2 illustrates such a network. In this game, the set of players is $N = \{sa, ab, bt, sc, ct, sd, dt\}$, the value of the coalition $C = \{sa, ab, bt\}$ is $v(N) = \min\{c_{sa}, c_{ab}, c_{bt}\} = 3$, the value of the coalition $D = \{sc, ct\}$ is $v(D) = \min\{c_{sc}, c_{ct}\} = 2$, and the value of the grand coalition N is $v(N) = 6$.

Several stability-related solution concepts for this class of games were studied by Granot and Granot [126]. Subsequently, Deng *et al.* [95] showed that computing the nucleolus in these games is easy for unit capacities, but hard in the general case.

One can also consider a variant of network flow games where the purpose is to send at least k units of flow from s to t: the value of a coalition is 1 if it can carry an s–t flow of size k and 0 otherwise. The resulting simple games are called *threshold network flow games*; they have been introduced by Bachrach and Rosenschein [34], and subsequently studied by Aziz *et al.* [19] and Resnick *et al.* [219]. For instance, for the network given in Figure 3.2, if $k = 2$, the coalition $\{sd, dt\}$ is losing, whereas the coalitions $\{sc, ct\}$ and $\{sa, ab, bt\}$ are winning.

Note that power indices such as the Shapley value or the Banzhaf index have a natural interpretation for this class of games, as they measure the importance of a given link in the network. Thus, they can be used as a rational basis on which to allocate maintenance and support budget. It would be therefore desirable to have an efficient algorithm for computing the power indices. Unfortunately, such an algorithm is unlikely to exist: Bachrach *et al.* [34] show that computing the Banzhaf power index in threshold network flow games is #P-complete. In fact, even deciding whether a player is a dummy, i.e., makes no contribution whatsoever, is computationally hard. On the positive side, Bachrach *et al.* [34] identify some special classes of graphs for which this problem is easy.

3.1.3   ASSIGNMENT AND MATCHING GAMES

In *assignment games* [234], agents are vertices of a weighted bipartite graph. The value of each coalition is the size of its maximum-weight induced matching. These games model simple market interactions, where the players belong to two groups, namely, "consumers" and "producers". Consumers and producers have to form exclusive bilateral contracts; for every possible partnership between a consumer and a producer, there is an associated value that indicates the potential profit from the relationship. These games can be cast as a special case of network flow games, but they have an additional structure that makes them more tractable. Granot and Granot [126] studied stability in assignment games, and showed some interesting relationships between the core, the kernel and the nucleolus in such games; their results were later used to design a polynomial-time algorithm for computing the nucleolus in assignment games [241]. *Matching games* [96] are a generalization of assignment games, where the graph is not required to be bipartite. The complexity of the core, the least core, and the nucleolus in these games has been studied by Kern and Paulusma [150].

3.1.4   MINIMUM COST SPANNING TREE GAMES

Minimum cost spanning tree games [51] differ from all the other games considered so far in that they are *cost-sharing games* rather than *profit-earning games*, i.e., the value of each coalition is non-positive. In these games, agents need to be connected to a certain service supplier, and form coalitions to split the cost of this service. Cost sharing games are discussed in more detail by Jain and Mahdian [146], where the authors view these games from a mechanism design perspective.

More specifically, a minimum cost spanning tree game is given by a set of agents N, a *supplier* s, and a complete graph $\mathcal{G} = (N \cup \{s\}, E)$ with edge costs c_{ij}. The value of a coalition $C \subseteq N$ is the cost of a minimum-cost spanning tree on $C \cup \{s\}$.

It is not hard to show that the core of a minimum cost spanning tree game is guaranteed to be non-empty. Indeed, we can fix a minimum-cost spanning tree of $\mathcal{G}$, and charge each agent the cost of the upstream link; a simple graph-theoretic argument shows that this cost allocation is stable [127]. On the other hand, Faigle *et al.* [112, 113] show that checking if a given outcome is in the core and computing the least core and the nucleolus is NP-hard.

3.1.5   FACILITY LOCATION GAMES

Just like spanning tree games, *facility location games* also belong to the class of cost-sharing games. However, they have a more complex combinatorial structure.

Specifically, in facility location games agents are located in vertices of a graph $\mathcal{G} = (V, E)$. In vertex v_i, one can open a *facility* at cost f_i. Each agent needs to be connected to some open facility. The cost of servicing a coalition C via a set of facilities F has two components: the opening costs and the connection costs, i.e.,

$$c(C, F) = \sum_{j \in F} f_j + \sum_{i \in C} d(i, F),$$

where $d(i, F)$ is the distance in $\mathcal{G}$ from i to the nearest facility in F.

The value of a coalition C can now be computed as the minimum cost of servicing C, over all possible sets of facilities:

$$v(C) = - \min_{F \subseteq V} c(C, F).$$

Stability in facility location games is studied in detail by Goemans and Skutella [125], who build on the seminal paper by Deng *et al.* [96] to show that such games have a non-empty core if and only if the integrality gap of a certain linear program is 0. The technique of Deng *et al.* [96] applies to several other classes of combinatorial optimization games; we refer the reader to a survey paper by Deng [94] for an overview.

3.2   COMPLETE REPRESENTATIONS

In this section, we survey several complete representation languages for characteristic function games that have been recently proposed in the literature.

3.2.1   MARGINAL CONTRIBUTION NETS

The *marginal contribution nets (MC-nets)* representation that was introduced by Ieong and Shoham [143] can be understood as an extension to the induced subgraph representation discussed in section 3.1.1. The basic idea behind marginal contribution nets is to represent the characteristic function of a game $G = (N, v)$ as a set of rules, of the form:

$$\text{pattern} \longrightarrow \text{value}.$$

Here, "pattern" is a Boolean formula over the variables $\{x_1, \ldots, x_n\}$, and "value" is a real number (which without loss of generality can be assumed to be integer). The Boolean variables $x_1, \ldots, x_n$ correspond to players $N = \{1, \ldots, n\}$. If the pattern of a rule in simply a conjunction of literals, i.e., variables or their negations, then we say the rule is *basic*. A basic rule is said to *apply* to a group of agents C if C contains all agents that appear unnegated in the pattern, and does not contain any of the agents that appear negated. For example, a rule with pattern $x_1 \wedge x_3 \wedge \neg x_7$ would apply to the coalitions $\{1, 3\}$ and $\{1, 3, 5\}$, but not to the coalitions $\{1\}$ or $\{1, 3, 7, 9\}$.

More generally, if ψ is an arbitrary Boolean formula over $\{x_1, \ldots, x_n\}$, we say that a coalition C *satisfies* ψ (and write $C \models \psi$) if ψ is satisfied by the truth assignment that sets $x_i = \text{true}$ if $i \in C$ and $x_i = \text{false}$ otherwise; a rule of the form $\psi \longrightarrow x$ *applies* to C if and only if $C \models \psi$.

Now, the value of a coalition is computed by summing the right-hand sides of all rules that apply to it. More formally, fix a collection of rules $R = \{r^1, \ldots, r^m\}$, where $r^i = \psi^i \longrightarrow x^i$ for each $i = 1, \ldots, m$ and each ψ^i is a Boolean formula over $\{x_1, \ldots, x_n\}$. The set R defines a coalitional game $G^R = (N, v^R)$ with a set of players $N = \{1, \ldots, n\}$ and a characteristic function v^R given by

$$v^R(C) = \sum_{r^i \in R \,:\, C \models \psi^i} x^i.$$

Let us consider an example. Suppose we have a ruleset R, containing the following rules:

$$\begin{aligned} x_1 \wedge x_2 &\longrightarrow 5 \\ x_2 &\longrightarrow 2 \\ x_3 &\longrightarrow 4 \\ x_2 \wedge \neg x_3 &\longrightarrow -2 \end{aligned}$$

Here, we have:

- $v^R(\{1\}) = 0$ (because no rules apply),

- $v^R(\{2\}) = 0$ (second and fourth rules),

- $v^R(\{3\}) = 4$ (third rule),

- $v^R(\{1, 2\}) = 5$ (first, second, and fourth rules),

- $v^R(\{1, 3\}) = 4$ (third rule),

- $v^R(\{2, 3\}) = 6$ (second and third rules), and

- $v^R(\{1, 2, 3\}) = 11$ (first, second, and third rules).

To see that MC-nets are universally expressive, observe that we can represent an arbitrary characteristic function game $G = (N, v)$ in this manner by introducing one rule for each coalition $C \subseteq N$: the left-hand side of this rule is given by $(\wedge_{i \in C} x_i) \wedge (\wedge_{i \notin C} \neg x_i)$, and its right-hand side is given by $v(C)$. Clearly, this rule applies to C only, and hence the value of C in the resulting game is exactly $v(C)$. Now, of course, this representation is not succinct. However, games where the value of a coalition is determined by the presence or absence of small groups of players often admit a fairly compact encoding in the MC-nets format. For instance, any induced subgraph game can be easily translated into a (basic) MC-net, with one rule for every edge.

What about the complexity of computing solution concepts under the MC-nets representation? For the Shapley value, we can use the same idea as we used for the induced subgraph representation, namely, decompose the game into smaller subgames. Indeed, it is easy to see that the game corresponding to the ruleset R can be viewed as the sum of the games corresponding to individual rules in R. Now, if these rules themselves have a simple structure, we can compute the Shapley values for each of the subgames, and then sum them up. Of course, if we allow arbitrary Boolean formulas in the pattern, this task is hopeless: a straightforward reduction from 3-SAT shows that deciding whether a player is a dummy (and hence deciding whether his Shapley value is 0) is computationally hard [107]. However, for basic MC-nets we can obtain an efficient algorithm in this manner [143].

To see this, consider a rule $\psi \longrightarrow x$ of a basic MC-net. In the corresponding game, all agents that appear unnegated in ψ are clearly symmetric; this is also the case for all agents that appear negated in ψ. Further, the value of each coalition is either x or 0, i.e., up to normalization, this is a simple game. Now, we need to apply some basic combinatorics: if φ contains p "positive" agents and s "negative" agents, a positive agent is pivotal for a random permutation if and only if it appears after all other positive agents and before all negative agents, i.e., with probability $\frac{(p-1)!s!}{(p+s)!}$. Thus, his Shapley value is exactly $\frac{(p-1)!s!}{(p+s)!}x$. For negative agents, the calculation is similar. Thus, we obtain a closed-form expression for calculating players' Shapley values in basic MC-nets. This easiness result can be extended to the so-called *read-once* MC-nets, where each variable appears at most once in each pattern, but the argument is somewhat more complicated [107].

Of course, for the core-related problems we cannot expect a positive result: indeed, these problems are already hard for induced subgraph games, and we have argued that MC-nets generalize induced subgraph games. However, Ieong and Shoham [143] show that if a natural graph that can be associated with an MC-net has bounded treewidth, then computing whether an outcome is in the core or deciding non-emptiness of the core becomes easy; these results were subsequently strengthened by Greco *et al.* [128].

3.2.2 SYNERGY COALITION GROUPS

Synergy Coalition Group (SCG) Representation [76] is a complete language for superadditive games that is obtained by "trimming down" the naive representation, i.e., one that lists all coalitions together with their values. It is based on the following idea. Suppose that a game $G = (N, v)$ is superadditive, and consider a coalition $C \subseteq N$. Then we have

$$v(C) \geq \max_{CS \in CS'_C} \sum_{C' \in CS} v(C'),$$ (3.1)

where CS'_C the set of all *non-trivial* coalition structures over C (i.e., all coalition structures except the one that consists of a single coalition C). Now, if the inequality (3.1) holds with equality, then there is no need to store the value of C as it can be computed from the values of the smaller coalitions. Therefore, we can represent G by listing the values of all coalitions of size 1 as well as the values of the coalitions for which there is a *synergy*—i.e., the inequality (3.1) is strict.

By construction, the SCG representation is complete. Moreover, it is succinct when there are only a few groups of agents that can collaborate productively. Further, it allows for an efficient procedure for checking whether an outcome is in the core: it can be shown that if an outcome is not in the core, then there is a "synergistic" coalition, i.e., one whose value is given explicitly in our representation, that can profitably deviate. However, the SCG representation has a major drawback: computing the value of a coalition may involve finding an optimal partition of the players into subcoalitions, and is therefore NP-hard.

3.2.3 SKILL-BASED REPRESENTATIONS

In many settings, the value of a coalition is derived directly from the skills possessed by the agents within it. A simple representation formalism that is based on this idea was proposed by Ohta *et al.* [198]. In this representation, there is a set of *skills* S, each agent $i \in N$ has a subset of the skills

$S_i \subseteq S$, and there is a function $f : 2^S \to \mathbb{R}$ such that the value of a coalition $C \subseteq N$ is given by $v(C) = f(\cup_{i \in C} S_i)$. Clearly, this representation is complete, as we can identify each agent i with a unique skill s_i and set $f(S') = v(\{i \mid s_i \in S'\})$ for any subset S' of the skill set. It is succinct when the performance of each coalition can be expressed in terms of a small number of skills possessed by the members of the coalition. Ohta *et al.* [198] discuss such representations in the context of anonymous environments, where agents can hide skills or split them among multiple identifiers (see section 7.2.1).

A related but richer representation was proposed by Bachrach and Rosenschein [33], who express the coalitional values in terms of skills and tasks. Specifically, in addition to the set of skills S, there is a set of *tasks* T, and every task $t \in T$ has a *skill requirement* $S^t \subseteq S$ and a payoff a. As before, each agent $i \in N$ has a set of skills $S_i \subseteq S$. It is assumed that a coalition $C \subseteq N$ can perform a task t if it has all skills that are required for t, i.e., if $S^t \subseteq \cup_{i \in C} S_i$; we will write $C \models t$ whenever this is the case. Finally, there is a task value function $F : 2^T \to \mathbb{R}$, which for every subset $T' \subseteq T$ of tasks specifies the payoff that can be obtained by a coalition that can perform all tasks in T'. The value of a coalition C is given by

$$v(C) = F(\{t \mid C \models t\}).$$

This representation is more compact than that of Ohta *et al.* [198] when the number of skills is large (so that the domain of the function f is very large), but the game can be described in terms of a small number of tasks, or if the function F can be encoded succinctly.

3.2.4   ALGEBRAIC DECISION DIAGRAMS

Algebraic decision diagrams (ADDs) [35] are a powerful formalism that is used to efficiently represent and analyze real-valued functions defined over the binary domain $\{0, 1\}^n$. In this formalism, a function $f(x_1, \dots, x_n)$ is represented by an acyclic directed graph whose internal nodes are labeled with variables from the set $\{x_1, \dots, x_n\}$; each node with label x_i has two outgoing edges that are labeled with 0 and 1, respectively. The graph has a single root (a vertex with indegree 0), and each of its terminal nodes (vertices with outdegree 0) is labeled with a real value. To compute the value of f at an input vector $(x_1, \dots, x_n) \in \{0, 1\}^n$, we start at the root and travel through the graph until we reach a terminal node; at the internal node with label x_i, we take the edge with label 0 if $x_i = 0$ and the edge with label 1 otherwise. We then output the label of the terminal node reached in this manner.

Recently, Aaditya *et al.* [1] and, independently, Ichimura *et al.* [142] proposed using ADDs to represent characteristic functions of coalitional games; a closely related approach was put forward by Bolus [56], who uses a precursor of algebraic decision diagrams known as *binary decision diagrams* to represent simple games. While Aaditya *et al.* use ADDs to directly encode the characteristic function of the game, Ichimura *et al.* combine ADDs with the synergy coalition groups representation discussed in section 3.2.2 to obtain a more succinct representation language.

It is easy to see that algebraic decision diagrams are universally expressive: any n-player characteristic function game can be represented by a complete binary tree with 2^n terminal nodes, and many interesting games admit a more succinct representation. For instance, Aaditya *et al.* show that any game with a constant number of agent *types* (two agents are said to have the same type if they are symmetric) can be succinctly represented as an ADD. An advantage of an ADD-based representation is that it admits polynomial time algorithms for many important problems, such as computing the Shapley value, the Banzhaf index, the value of the least core, and the cost of stability, as well as checking whether the core is non-empty and whether a given outcome belongs to the core.

3.3   ORACLE REPRESENTATION

In this section, we consider the setting where the characteristic function of a game is given by a black box, or an oracle, i.e., we have access to an efficient procedure that, given a coalition, outputs its value.

While designing efficient algorithms in this very general model may appear to be a hopeless task, some of the results derived in chapter 2 can be interpreted in this way: for instance, for simple games, Theorem 2.26 provides an efficient algorithm for checking non-emptiness of the core/verifying if a given outcome is in the core in the oracle model, while Theorem 2.27 describes an algorithm for constructing an outcome in the core of a convex game in the oracle model. Moreover, the results of section 2.2.3.6 can also be interpreted in this vein: if we have an oracle not just for the characteristic function, but also for checking if a given outcome is in the core, we can decide whether the core is non-empty, and with a slightly more powerful oracle we can also compute the value of the least core and the cost of stability.

However, the hardness results discussed earlier in this chapter (as well as in chapter 4) outline the limitations of this approach. Indeed, any succinct representation of a coalitional game can be used to construct an oracle for this game. Therefore, if a certain problem (such as computing the Shapley value or deciding non-emptiness of the core) has been shown to be hard for a specific representation, the same hardness result holds in the oracle model. For instance, since computing the Shapley value is #P-hard for weighted voting games (see chapter 4), it is also hard in the oracle model; similarly, since checking whether an outcome is in the core is coNP-hard for induced subgraph games (see section 3.1.1), it is hard in the oracle model as well.

Nevertheless, it turns out that, for many stability-related solution concepts, one can prove complexity upper bounds in the oracle model that match the existing hardness results in much simpler models. Specifically, Greco *et al.* [128] show that under the oracle representation checking whether the core is non-empty is in coNP, checking if an outcome is in the kernel is in Δ_2^p, and checking if an outcome is in the bargaining set is in Π_2^p, thus matching the hardness results for induced subgraph games. In a subsequent paper [129], Greco *et al.* analyze the complexity of stability-related solution concepts in the presence of coalition structures.

CHAPTER 4

Weighted Voting Games

Weighted voting games form one of the simplest useful classes of coalitional games. These games can be used to model settings where each player has a certain amount of a given resource (say, time, money, or manpower), and there is a goal that can be reached by any coalition that possesses a sufficient amount of this resource; a typical example is parliamentary voting, where players are parties, and the resource of each party is the number of votes it controls. In this chapter, we study weighted voting games in detail. Their simplicity enables us to fully describe several algorithms for computing solution concepts for this class of games, or to give computational hardness proofs. In other words, we use weighted voting games as a case study in order to illustrate how one can go about computing a given solution concept. Towards the end of the chapter, we will also discuss a generalization of weighted voting games, known as vector weighted voting games.

4.1 DEFINITION AND EXAMPLES

We start by giving a formal definition of weighted voting games.

Definition 4.1 A *weighted voting game* G with a set of players $N = \{1, \ldots, n\}$ is given by a list of *weights* $\mathbf{w} = (w_1, \ldots, w_n) \in \mathbb{R}^n$ and a *quota* $q \in \mathbb{R}$; we will write $G = [N; \mathbf{w}; q]$. Its characteristic function $v : 2^N \to \{0, 1\}$ is given by

$$v(C) = \begin{cases} 1 & \text{if } \sum_{i \in C} w_i \geq q \\ 0 & \text{otherwise.} \end{cases}$$

We will follow the convention introduced in Definition 2.4 and write $w(C)$ to denote the total weight of a coalition C, i.e., we set $w(C) = \sum_{i \in C} w_i$. Also, we set $w_{\max} = \max_{i \in N} w_i$.

It is usually assumed that all weights and the quota are non-negative; in what follows, we will make this assumption as well. Under this assumption, weighted voting games are monotone and, moreover, belong to the class of simple games. In section 4.4 we will see that, in a sense, the converse is also true: any simple game can be represented as a combination of weighted voting games. Also, it is standard to assume that $0 < q \leq w(N)$; this condition ensures that the empty coalition is losing and the grand coalition is winning.

Equivalence

Two weighted voting games may have different weights and quotas, yet encode the same simple game, i.e., have exactly the same set of winning coalitions. For instance, this is the case if one game is obtained from the other by scaling up all weights and the quota by the same factor. Another example is provided by the games $G = [\{1, 2\}; (3, 3); 5]$ and $G' = [\{1, 2\}; (1, 1); 2]$: in both games, the only winning coalition is the grand coalition. If this is the case, we will say that the two games are *equivalent*. Formally, two weighted voting games $G = [N; \mathbf{w}; q]$ and $G = [N'; \mathbf{w}'; q']$ are said to be equivalent if $N = N'$ and for every $C \subseteq N$ it holds that $w(C) \geq q$ if and only if $w'(C) \geq q'$. The notion of equivalence naturally extends to other representations of simple games, and, in particular, to vector weighted voting games, which are the subject of section 4.4.

Integrality of Weights and Quota

Definition 4.1 allows for arbitrary real weights. However, it is not clear how to efficiently store and manipulate such weights, which presents a difficulty from the algorithmic point of view. Fortunately, it turns out that any weighted voting game is equivalent to a game with fairly small integer weights. More precisely, we have the following theorem, which follows from results on linear threshold functions [188].

Theorem 4.2   For any weighted voting game $G = [N; \mathbf{w}; q]$ with $|N| = n$, there exists an equivalent weighted voting game $G' = [N; \mathbf{w}'; q']$ with $\mathbf{w}' = (w'_1, \ldots, w'_n)$ such that all $w'_i, i = 1, \ldots, n$, and q' are non-negative integers, and $w'_i = O(2^{n \log n})$ for all $i = 1, \ldots, n$.

Therefore, in what follows, we can assume without loss of generality that all weights and the quota are integers given in binary. We remark that, even though the entries of the weight vector $\mathbf{w}'$ are exponential in n, they can be represented using $O(n \log n)$ bits, i.e., a weighted voting game with n players can be described using $\text{poly}(n)$ bits.

Applications

Weighted voting games describe many real-world situations. In particular, they are very well suited to model coalition formation in legislative bodies. In more detail, each party in a parliament can be associated with a player in a weighted voting game; the weight of the player is given by the total number of the representatives of this party. The quota corresponds to the number of votes needed to pass a bill; while in most cases a simple majority $q = \lfloor w(N)/2 \rfloor + 1$ suffices, in some circumstances a bill can only be passed if it is supported by two thirds of the legislators (i.e., $q = 2w(N)/3$), or even the entire legislative body ($q = w(N)$). Another example of a weighted voting game is shareholder voting: the weight of each voter is proportional to the number of shares she holds. A weighted voting game also arises in a setting where there is a task that requires q hours of work, and there is a group of agents $N = \{1, \ldots, n\}$ such that each agent i can dedicate w_i of his time to this task. Finally, the variant of the ice cream game where there is only one tub size (normalized to 1) is a weighted voting game as well.

4.2   DUMMIES AND VETO PLAYERS

We have observed that weighted voting games are simple games. In chapter 2, we identified two special categories of players in simple games, which play an important role in the analysis of such games, namely, dummies and veto players. It is therefore natural to ask if we can efficiently decide whether a given player in a weighted voting game belongs to one of these categories. We will first state the respective computational problems formally.

Name VETO.

Instance A weighted voting game $G = [N; \mathbf{w}; q]$ and a player $i \in N$.

Question Is i a veto player in G?

Name DUMMY.

Instance A weighted voting game $G = [N; \mathbf{w}; q]$ and a player $i \in N$.

Question Is i a dummy player in G?

As argued in chapter 2, finding the veto players is easy for any simple game with an efficiently computable characteristic function: given a player i, it is enough to check whether the coalition $N \setminus \{i\}$ is winning. For weighted voting games, this means checking whether $w(N \setminus \{i\}) \geq q$. This immediately implies the following result.

Proposition 4.3   VETO *is polynomial-time solvable.*

In contrast, identifying the dummy players turns out to be difficult. To show this, we provide a reduction from the classic Partition problem [122, p.223], defined as follows.

Name PARTITION.

Instance Positive integers $a_1, \ldots, a_k$, K such that $\sum_{i=1}^{k} a_i = 2K$.

Question Is there a subset of indices J such that $\sum_{i \in J} a_i = K$?

Theorem 4.4   DUMMY is *coNP-complete.*

Proof. If a player i is not a dummy player, this can be proved by exhibiting a coalition C such that $w(C) < q$, $w(C \cup \{i\}) \geq q$. Thus, DUMMY is in coNP.

To show coNP-hardness, we will transform an instance $I = (a_1, \ldots, a_k, K)$ of PARTITION into a weighted voting game $G = [N; \mathbf{w}; q]$, which is constructed as follows. We set $N = \{1, \ldots, k, k+1\}$, $w_i = 2a_i$ for $i = 1, \ldots, k$, $w_{k+1} = 1$, and $q = 2K + 1$. It is easy to see that I is a "yes"-instance of PARTITION if and only if player $k+1$ is not a dummy in G. Indeed, suppose that there is a subset of indices J such that $\sum_{i \in J} a_i = K$. Then we have $w(J) = 2K$, $w(J \cup \{k+1\}) = 2K + 1 = q$, i.e., $k+1$ is pivotal for $J \cup \{k+1\}$. Conversely, suppose that $k+1$

is pivotal for some coalition C. Then we have $w(C \setminus \{k+1\}) < q, w(C) \geq q$. Since $w_{k+1} = 1$ and all weights are integer, this means that $w(C \setminus \{k+1\}) = 2K$, i.e., $\sum_{i \in C \setminus \{k+1\}} a_i = K$.

We conclude that I is a "yes"-instance of PARTITION if and only if (G, k) is a "no"-instance of DUMMY. Therefore, DUMMY is coNP-hard. □

Our proof that DUMMY is coNP-hard is very simple. However, later we will see that the same proof idea can be used to obtain many other hardness results for weighted voting games.

Another important observation is that in our hardness reduction, the weights are derived from the numbers in the instance of Partition. Thus, our hardness result is relevant only if the weights are fairly large. Put differently, we have shown that Dummy is unlikely to admit an algorithm that runs in time polynomial in the input size, i.e., poly$(n, \log w_{\max})$. Now, in some applications of weighted voting games, such as shareholder voting, the weights (i.e., numbers of shares held by individual shareholders) can be huge. However, in parliamentary voting the weight of each party is usually fairly small: for instance, at the time of writing the UK House of Commons has 650 seats, and the Israeli Knesset has 120 seats. In such settings, we might be satisfied with a *pseudopolynomial* algorithm, i.e., an algorithm that runs in time poly$(n, w_{\max})$, or, equivalently, runs in polynomial time if all numbers in the input are given in unary. It turns out that Dummy admits such an algorithm. This will follow from Theorem 4.8 in the next section, which shows that each player's Shapley value in a weighted voting game can be computed in pseudopolynomial time. Indeed, since weighted voting games are monotone, a player is a dummy in a weighted voting game if and only if her Shapley value is 0. Therefore, an algorithm for computing players' Shapley values can be used to identify dummies.

4.2.1   POWER AND WEIGHT

We have already argued that an agent's power in a coalitional game can be measured by her Shapley value or her Banzhaf index. If the game in question is a weighted voting game, one may expect that an agent's Shapley value is closely related to her weight. This is not entirely incorrect: it is not hard to show that power is monotone in weight, i.e., for any weighted voting game $G = [N; \mathbf{w}; q]$ and any two players $i, j \in N$ we have $\varphi_i(G) \leq \varphi_j(G)$ if and only if $w_i \leq w_j$. However, two agents may have identical voting power even if their weights differ considerably.

Example 4.5   After the May 2010 elections in the UK, the Conservative Party had 307 seats, the Labour Party had 258 seats, the Liberal Democrats (LibDems) had 57 seats, and all other parties shared the remaining 28 seats (with the most powerful of them getting 8 seats). It is easy to see that in this weighted voting game there are two two-party coalitions (Conservatives+Labour and Conservatives+LibDems) that can get a majority of seats. Moreover, if Labour or LibDems want to form a coalition that does not include Conservatives, they need each other (as well as a few minor parties). Thus, Labour and LibDems have the same Shapley value, despite being vastly different in size.

The phenomenon illustrated in Example 4.5 has been observed in many decision-making bodies. It explains why major parties often end up making concessions to smaller parties in order

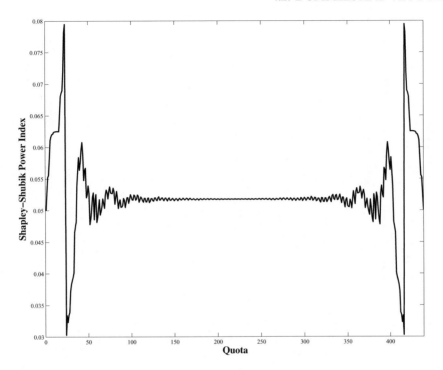

Figure 4.1: The Shapley value of player 10 (weight 23) for weight vector (1, 2, 4, 5, 16, 17, 20, 21, 21, 23, 24, 24, 27, 28, 28, 33, 33, 36, 36, 40).

to form a winning coalition: the small parties may wield substantial voting power. Example 4.5 also indicates that to determine an agent's power, we have to take into account the distribution of the other players' weights as well as the quota. In particular, if we keep the weights fixed, but alter the quota, an agent's power can change considerably.

Example 4.6   Consider a weighted voting game with two players of weight 4 and two players of weight 1. If the quota is set to 10, the only winning coalition is the grand coalition, so each player's Shapley value is 1/4. If the quota is set to 8, the smaller players are dummies, so their Shapley value is 0. Finally, if the quota is set to 5, a player of weight 1 is pivotal only if it appears in the second position, and a player of weight 4 appears in the first position. There are four permutations that satisfy this condition, so for $q = 5$ the Shapley value of each of the smaller players is 1/6.

The role of the quota in determining the agents' power in weighted voting games was studied in detail by Zuckerman *et al.* [272] and then by Zick *et al.* [271]. The graphs in Figures 4.1 and 4.2 (from [271]) illustrate the changes in a player's power as the quota varies from 1 to $w(N)$.

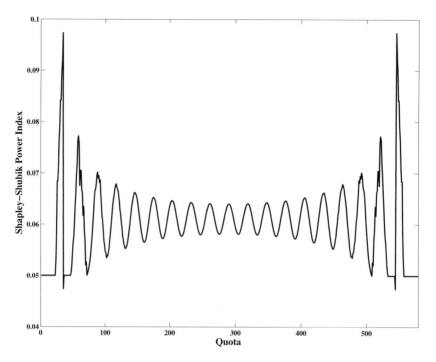

Figure 4.2: The Shapley value of player 19 (weight 35) for weight vector $(23, 24, 24, 25, 25, 25, 25, 27, 28, 28, 29, 30, 30, 32, 32, 33, 34, 34, 35, 36)$.

Observant readers will notice that both graphs are centrally symmetric. We will now show that this is true for all weighted voting games; this property is known as *self-duality* [117].

Proposition 4.7   *Given a weighted voting game $G = [N; \mathbf{w}; q]$, let $G' = [N; \mathbf{w}; w(N) + 1 - q]$. Then for any player $i \in N$ it holds that $\varphi_i(G) = \varphi_i(G')$.*

Proof. Suppose i is pivotal for a permutation π in the game G, i.e.,

$$w(S_\pi(i)) < q, \quad w(S_\pi(i) \cup \{i\}) \geq q.$$

Let π' be the permutation obtained by reversing π. Then we have

$$w(S_{\pi'}(i)) = w(N) - w(S_\pi(i)) - w_i \leq w(N) - q < w(N) - q + 1,$$
$$w(S_{\pi'}(i) \cup \{i\}) = w(N) - w(S_\pi(i)) > w(N) - q \geq w(N) - q + 1.$$

Hence, i is pivotal for the permutation π' in the game G'. By symmetry, the converse is also true: if i is pivotal for π' in G', then i is pivotal for π in G. Thus, we have established a bijection between the set of permutations that i is pivotal for in G and the set of permutations that i is pivotal for in G'. Consequently, $\varphi_i(G) = \varphi_i(G')$. □

Another interesting property of the graphs in Figures 4.1 and 4.2 is that they have a peak at $q = w_i$ (and, by symmetry, at $q = w(N) + 1 - w_i$). Zick *et al.* [271] showed that this holds for all players in all weighted voting games. Similarly to the proof of Proposition 4.7, the argument is based on counting permutations that i is pivotal for, for different values of q. The experiments described by Zick *et al.* also show that, in a significant fraction of randomly generated games, setting $q = w_i + 1$ *minimizes* player i's Shapley value, i.e., a very small change of the quota may lead to a sharp change in the player's power.

4.2.2   COMPUTING THE POWER INDICES

In order to use the Shapley value and the Banzhaf index to measure the agents' power in weighted voting games, we would like to have an efficient algorithm for computing these indices. However, such an algorithm is unlikely to exist: Theorem 4.4, combined with the dummy player axiom and the fact that weighted voting games are monotone, directly implies that checking whether an agent's Shapley value is 0 is coNP-hard. In fact, computing the Shapley value and the Banzhaf index is known to be #P-complete [97, 115, 178, 210]. However, just as in the case of DUMMY, these hardness results are only relevant when the weights can be assumed to be large. For small weights both the Shapley value and the Banzhaf index can be computed by a pseudopolynomial algorithm, as shown by Matsui and Matsui [177].

Theorem 4.8   Given an n-player weighted voting game $G = [N; \mathbf{w}; q]$ and a player $i \in N$, we can compute $\beta_i(G)$ and $\varphi_i(G)$ in time $O(n^2 w_{\max})$ and $O(n^3 w_{\max})$, respectively.

Proof. We will first describe the algorithm for the Shapley value; later, we will explain how to simplify it for the Banzhaf index.

By renumbering the players if necessary, we can assume that $i = n$, i.e., our goal is to compute the Shapley value of the last player. We can assume without loss of generality that $w_n > 0$, since otherwise we clearly have $\varphi_n(G) = \beta_n(G) = 0$. Observe first that it suffices to determine, for each $s = 0, \dots, n - 1$, the number N_s of s-element subsets of $N \setminus \{n\}$ that have weight $W \in \{q - w_n, \dots, q - 1\}$. Indeed, whenever i is pivotal for a coalition C, $|C| = s + 1$, it is pivotal for all permutations in which the agents in $C \setminus \{i\}$ appear in the first s positions, followed by i; there are exactly $s!(n - s - 1)!$ such permutations (where we use the convention that $0! = 1$). Therefore, the formula for the Shapley value can be rewritten as

$$\varphi_i(G) = \frac{1}{n!} \sum_{s=0}^{n-1} s!(n - s - 1)! N_s. \tag{4.1}$$

To compute N_s, we use dynamic programming. Specifically, we define $X[j, W, s]$ to be the number of s-element subsets of $\{1, \dots, j\}$ that have weight W; here, j ranges from 1 to $n - 1$, s

ranges from 0 to $n - 1$, and W ranges from 0 to $w(N)$. For $s = 0$, $j = 1, \ldots, n - 1$, we have

$$X[j, W, 0] = \begin{cases} 1 & \text{if } W = 0 \\ 0 & \text{otherwise.} \end{cases}$$

Further, for $j = 1, s = 1, \ldots, n - 1$ we have

$$X[1, W, s] = \begin{cases} 1 & \text{if } W = w_1 \text{ and } s = 1 \\ 0 & \text{otherwise.} \end{cases}$$

Now, having computed the values $X[j', W', s']$ for all $j' < j$, all $W' = 0, \ldots, w(N)$, and all $s' = 0, \ldots, n - 1$, we can compute $X[j, W, s]$ for $W = 0 \ldots, w(N)$ and $s = 1, \ldots, n - 1$ as follows:

$$X[j, W, s] = X[j - 1, W, s] + X[j - 1, W - w_j, s - 1]$$

In the equation above, the first term counts the number of subsets that have weight W and size s and do not contain j, whereas the second term counts the number of subsets of this weight and size that do contain j.

Thus, we can inductively compute $X[n - 1, W, s]$ for all $W = 0, \ldots, w(N)$ and all $s = 0, \ldots, n - 1$. Now, $N_s, s = 0, \ldots, n - 1$, can be computed as

$$N_s = X[n - 1, q - w_n, s] + \cdots + X[n - 1, q - 1, s].$$

By substituting this expression into equation (4.1), we can compute the Shapley value of player n.

The running time of this algorithm is dominated by the time needed to fill out the table $X[j, W, s]$. The size of this table can be bounded by $n \times n w_{max} \times n$, and each of its entries can be computed in $O(1)$ steps, which proves our bound on the running time.

For the Banzhaf index, the dynamic program can be simplified by omitting the third index, s: indeed, to compute the Banzhaf index, we simply need to know how many subsets of $N \setminus \{n\}$ have weight that is at least $q - w_n$ and at most $q - 1$. This allows us to reduce the running time from $O(n^3 w_{max})$ to $O(n^2 w_{max})$.   □

When the weights are large, Theorem 4.8 is not very useful, and we may want to resort to heuristics and/or approximation algorithms for computing the power indices. Now, Theorem 4.4 implies that, unless P=NP, no efficient algorithm can approximate the Shapley value within a constant factor on all instances. However, it does not rule out the existence of randomized algorithms that are *probably approximately correct (PAC)*, i.e., produce a good solution with high probability.

One such algorithm for this problem is given by Bachrach *et al.* [29]. The main idea of this algorithm, which dates back to Mann and Shapley [171], is to randomly sample a coalition and check whether the given player is pivotal for it. It is not hard to see that the fraction of coalitions for which i is pivotal provides an unbiased estimator of i's Banzhaf index. Bachrach *et al.* [29] show that, by averaging over poly$(n, \ln 1/\delta, 1/\varepsilon)$ samples, we can obtain an estimate that is within a distance ε

from i's true Banzhaf index with probability at least $1 - \delta$; this approach can be generalized to the Shapley value.

Several other papers consider the problem of computing the power indices, either approximately or exactly; an (incomplete) list includes [45, 116, 164, 172, 180, 201].

4.2.3   PARADOXES OF POWER

An agent's Shapley value and his Banzhaf index may behave in an unexpected way if we modify the game. For example, we might naively expect that adding players to a game would reduce the power of players already present in the game, but this is not necessarily the case: when a new player joins the game, the power of some existing players may in fact increase.

Example 4.9   Consider a weighted voting game $G = [\{1, 2, 3\}; (2, 2, 1); 4]$. Clearly, player 3 is a dummy in this game, so $\varphi_3(G) = 0$. Now, suppose that a new player with weight 1 joins this game. In the resulting game G', player 3 is pivotal for the coalition consisting of himself, the new player and one of the other two players, so $\varphi_3(G') > 0$.

Another interesting observation is that, when a player i in a game G splits into two different players, i.e., distributes his weight between two identities i' and i'', the sum of the new identities' Shapley values in the resulting game can be considerably different from i's Shapley value in the original game.

Example 4.10   Consider an n-player weighted voting game $G = [N; \mathbf{w}; q]$ with $\mathbf{w} = (2, 1, \ldots, 1)$ and $q = n + 1$. In this game the only winning coalition is the grand coalition, so $\varphi_i(G) = \frac{1}{n}$ for all $i = 1, \ldots, n$. Now, suppose that player 1 decides to split into two unit-weight players $1'$ and $1''$. In the resulting game $G' = [N \setminus \{1\} \cup \{1', 1''\}; (1, \ldots, 1); n + 1]$ all players are symmetric, and therefore have equal Shapley value, namely, $\frac{1}{n+1}$. Thus, the joint power of the two new identities of player 1 is $\frac{2}{n+1}$, i.e., almost twice his original power.

However, weight-splitting may also lead to a reduction in power. To see this, consider an n-player weighted voting game $G = [N; \mathbf{w}; q]$ with $\mathbf{w} = (2, 2, \ldots, 2)$ and $q = 2n - 1$. By symmetry, we have $\varphi_i(G) = \frac{1}{n}$ for all $i = 1, \ldots, n$. However, it can be shown that if player 1 splits into two unit-weight players $1'$ and $1''$, the sum of their Shapley values in the new game $G' = [N \setminus \{1\} \cup \{1', 1''\}; (1, 1, 2, \ldots, 2); 2n - 1]$ is only $\frac{2}{n(n+1)}$; the proof of this fact is left as an exercise for the reader. Thus, weight-splitting lowers the agent's power by a factor of $(n + 1)/2$.

The phenomena illustrated in Examples 4.9 and 4.10 are known as *the paradox of new members* and *the paradox of size*, respectively. There are several other forms of counterintuitive behavior exhibited by the power indices; jointly, they are referred to as *the paradoxes of power*. These paradoxes are discussed by Felsenthal and Machover [117] and subsequently by Laruelle and Valenciano [162] (see also the references therein). The paradox of size is studied in detail by Aziz *et al.* [18], who show that the games described in Example 4.10 exhibit the strongest form of this paradox possible: in an

n-player weighted voting game, splitting into two identities can increase (respectively, decrease) an agent's Shapley value by at most a factor of $2n/(n + 1)$ (respectively, $(n + 1)/2$). Aziz *et al.* [18] also show that deciding whether a given player can split so as to increase his power is NP-hard.

4.3   STABILITY IN WEIGHTED VOTING GAMES

The complexity of core-related questions for weighted voting games strongly depends on whether we allow outcomes in which agents do not form the grand coalition. Traditionally, for simple games (and hence for weighted voting games) the answer to this question is negative, i.e., the only admissible outcomes are the ones where the grand coalition forms. In this case, Theorem 2.26 immediately implies that it is easy to check whether a given weighted voting game has a non-empty core. Indeed, it suffices to check if any of the players is a veto player, and, by Proposition 4.3, this check can be performed in polynomial time. Similarly, to check whether an outcome $\mathbf{x}$ (i.e., a payoff vector for the grand coalition) is in the core, it is enough to verify that $x_i = 0$ for any player i that is not a veto player.

However, weighted voting games are often not superadditive. In particular, if $q < w(N)/2$, the players may be able to form two or more disjoint winning coalitions. While such values of q do not make sense in voting scenarios, they can model the multiagent task allocation setting discussed in the beginning of the chapter: several groups of agents may independently embark on different tasks. Therefore, weighted voting games in which players may form non-trivial coalition structures are of interest to multiagent systems researchers. We will now present several results for such games; our exposition is based on the results of Elkind *et al.* [105]. To emphasize that we allow outcomes where the grand coalition does not form, we will refer to the set of outcomes of the form $(CS, \mathbf{x})$ that are stable against coalitional deviations as the *CS-core* of a weighted voting game.

First, we observe that weighted voting games with coalition structures admit a stable outcome whenever the agents can be partitioned into coalitions of weight exactly q.

Proposition 4.11   *Let $G = [N; \mathbf{w}; q]$ be a weighted voting game. Suppose that there exists a coalition structure $CS = \{C_1, \ldots, C_k\}$ over N such that $w(C_j) = q$ for all $j = 1, \ldots, k$. Then the CS-core of G is non-empty.*

Proof.   Set $x_i = w_i/q$ for any $i \in N$, and consider the outcome $(CS, \mathbf{x})$. We have $x(C_j) = w(C_j)/q = 1$, so $\mathbf{x}$ is a payoff vector for CS. Further, for any coalition C with $w(C) \geq q$ we have $x(C) = w(C)/q \geq 1$, so no coalition has a profitable deviation from $(CS, \mathbf{x})$. Thus, $(CS, \mathbf{x})$ is in the CS-core of G.   □

However, it turns out that deciding whether a given weighted voting game admits a stable coalition structure is hard. We will now formalize this computational problem, and present the hardness reduction given in [105].

Name CS-CORE.

Instance A weighted voting game $G = [N; \mathbf{w}; q]$, where agents may form coalition structures.

Question Is the CS-core of G non-empty?

Theorem 4.12   CS-CORE is *NP*-hard.

Proof. Given an instance $I = (a_1, \ldots, a_k, K)$ of PARTITION, we construct a weighted voting game $G = [N; \mathbf{w}; q]$ with $N = \{1, \ldots, k\}$, $w_i = a_i$ for $i = 1, \ldots, k$, and $q = K$. We can assume without loss of generality that $a_i \leq K$ for all $i = 1, \ldots, k$: if this is not the case, I is obviously a "no"-instance of PARTITION.

Suppose first that I is a "yes"-instance of PARTITION, i.e., there exists a subset of indices J such that $\sum_{i \in J} a_i = K$. Then $w(J) = w(N \setminus J) = q$, so by Proposition 4.11 G has a non-empty CS-core.

Conversely, suppose that I is a "no"-instance of PARTITION. It is easy to see that in this case any coalition structure contains at most one winning coalition, so the players' total payoff is at most 1. Now, any coalition structure that contains no winning coalitions is obviously unstable: the players can deviate by forming the grand coalition, which has value 1. On the other hand, consider a coalition structure CS that contains exactly one winning coalition C, and let $\mathbf{x}$ be the corresponding payoff vector. At least one player $i \in C$ must receive a strictly positive payoff $x_i > 0$. Thus, the total payoff obtained by the players in $N \setminus \{i\}$ is strictly less than 1. However, since $w_i \leq K$, we have $w(N \setminus \{i\}) \geq q$, so the coalition $N \setminus \{i\}$ has value 1, and therefore $(CS, \mathbf{x})$ is not in the CS-core. $\square$

We remark that it is not clear whether CS-CORE is in NP: after we guess a putative core outcome, we still need to make exponentially many checks to verify that it is indeed stable. Another difficulty is that, when guessing an outcome, we have to specify the entries of the payoff vector $\mathbf{x}$, and it is not immediately clear that they can be described using poly$(n, \log w_{max})$ bits. Fortunately, the latter issue can be resolved: Elkind *et al.* [105] show that if a weighted voting game G has a non-empty CS-core, then its CS-core contains an outcome that can be specified by poly$(n, \log w_{max})$ bits. This means that CS-CORE is in Σ_2^p. A recent result of Greco *et al.* [129] strengthens this upper bound to Δ_2^p; in fact, the bound of Greco *et al.* applies to any non-superadditive game with an efficiently computable characteristic function. However, at present the exact complexity of CS-CORE is not known. Further, it is not known if this problem remains hard when the weights are given in unary.

A related problem is that of checking whether a given outcome $(CS, \mathbf{x})$ is in the CS-core (assuming that the entries of $\mathbf{x}$ are rational numbers given in binary). This problem can be shown to be coNP-complete, but solvable in polynomial time if the weights are given in unary [105]. The hardness proof is similar to that of Theorem 4.12. To prove the easiness result, we reduce this problem to the classic KNAPSACK problem, defined as follows.

Name KNAPSACK.

Instance A list of k items with non-negative integer utilities $u_1, \ldots, u_k$ and non-negative integer sizes $s_1, \ldots, s_k$, the knapsack size S and the target utility U.

Question Is there a subset of indices $J \subseteq \{1, \ldots, k\}$ such that $\sum_{i \in J} s_i \leq S$ and $\sum_{i \in J} u_i \geq U$?

This problem is known to be NP-complete [122]; however, it can be solved in polynomial time if either sizes or utilities are given in unary. More formally, it admits an algorithm whose running time is

$$\min\{\text{poly}(k, s_{\max}, \log u_{\max}), \text{poly}(k, u_{\max}, \log s_{\max})\},$$

where $s_{\max} = \max_{i=1,\ldots,k} s_i$, $u_{\max} = \max_{i=1,\ldots,k} u_i$.

Now, to prove that a given outcome $(CS, \mathbf{x})$ is not stable, it suffices to find a coalition C with $w(C) \geq q$, $x(C) < 1$, i.e., to solve an instance of KNAPSACK problem with sizes given by $\mathbf{x}$ and utilities given by $\mathbf{w}$ (we will have to scale up the entries of $\mathbf{x}$ to ensure that they are integer, but this does not change the instance description size). Unary weights correspond to unary utilities in the KNAPSACK instance; therefore, our problem is easy for unary weights.

4.3.1 THE LEAST CORE, THE COST OF STABILITY, AND THE NUCLEOLUS

We have already seen that if weights cannot be assumed to be small, many solution concepts for weighted voting games are hard to compute. This is also the case for all three solution concepts that are the subject of this section, namely, the least core, the cost of stability, and the nucleolus[1]. More precisely, for the least core we consider the following decision problem:

Name EPSILONCORE.

Instance A weighted voting game $G = [N; \mathbf{w}; q]$, and a rational value $\varepsilon \geq 0$.

Question Is the ε-core of G non-empty?

Clearly, (G, ε) is a "yes"-instance of EPSILONCORE if and only if the value of the least core of G is at most ε, i.e., this problem captures the complexity of computing the value of the least core. Elkind *et al.* [106] showed that EPSILONCORE is coNP-hard; the proof is similar to the proofs of Theorems 4.4 and 4.12, and is left as an exercise. It is not clear, however, if EPSILONCORE is in coNP; indeed, the best complexity upper bound that is known for this problem is Σ_2^p (we can first guess a solution and then verify that no coalition can gain more than ε by deviating).

Besides computing the value of the least core, one may want to check if a given outcome is in the least core or in the ε-core for a given value of ε; another natural computational problem is constructing an outcome in the least core/ε-core (for a given value of ε). These problems, too, can be shown to be computationally hard, using essentially the same construction. On the other hand,

---

[1]These solution concepts have been defined for superadditive games only, whereas weighted voting games are not necessarily superadditive. Nevertheless, in the rest of this section, we will follow the approach usually taken in the weighted voting games literature and treat weighted voting games as superadditive games, i.e., assume that agents always form the grand coalition.

each of these problems admits a pseudopolynomial algorithm. We provide a formal proof for the problem of computing the value of the least core; however, it is easy to see that our argument extends to the remaining four problems.

Theorem 4.13   The value of the least core can be computed in time $\mathrm{poly}(n, w_{\max})$.

Proof. Suppose that we are given a weighted voting game $G = [N; \mathbf{w}; q]$ with $|N| = n$. For this game, the linear program for the least core given by (2.2) takes the following form:

$$
\begin{aligned}
\min \quad & \varepsilon \\
& x_i \;\geq\; 0 \quad \text{for each } i \in N \\
& \sum_{i \in N} x_i \;=\; 1 \\
& \sum_{i \in C} x_i \;\geq\; 1 - \varepsilon \quad \text{for each } C \subseteq N \text{ such that } \sum_{i \in C} w_i \geq q
\end{aligned}
\tag{4.2}
$$

This linear program attempts to choose an imputation $\mathbf{x}$ so as to minimize the deficit ε of the worst-off winning coalition. Any solution to this linear program is a vector of the form $(x_1, \ldots, x_n, \varepsilon)$. Moreover, if $(\mathbf{x}, \varepsilon)$ is an optimal solution, then the payoff vector $(x_1, \ldots, x_n)$ is in the least core, and the value of the least core is ε.

As argued in section 2.2.3, to solve this linear program in pseudopolynomial time, it suffices to design a pseudopolynomial time separation oracle for it. Recall that a separation oracle for a linear program is an algorithm that, given a putative feasible solution, checks whether it is indeed feasible, and if not, outputs a violated constraint [131]. In our case, this means that we need an algorithm that given a vector $(x_1, \ldots, x_n, \varepsilon)$, checks if there is a winning coalition C such that $\sum_{i \in C} x_i < 1 - \varepsilon$. The running time of this algorithm has to be polynomial in n, $||(\mathbf{x}, \varepsilon)||$ and $w_{\max}$, where $||(\mathbf{x}, \varepsilon)||$ denotes the number of bits needed to specify $(x_1, \ldots, x_n, \varepsilon)$ (the coordinates of this vector can be assumed to be rational).

Our algorithm proceeds as follows. First, it sets $p = 1 - \varepsilon$ and finds the smallest constant R such that $p' = pR$ and all $x_i' = x_i R$, $i = 1, \ldots, n$, are integer. Then it constructs an instance I of KNAPSACK by setting $s_i = x_i'$, $u_i = w_i$ for $i = 1, \ldots, n$, $S = p' - 1$, $U = q$, and runs the pseudopolynomial algorithm for KNAPSACK on I. Clearly, if I is a "no"-instance, then all constraints are satisfied. On the other hand, if I is a "yes"-instance, we can identify a subset of items J that witnesses this (i.e., has $s(J) \leq S$, $u(J) \geq U$), by using the standard dynamic programming techniques. This subset corresponds to a violated constraint. Finally, observe that

$$
\log s_{\max} = \log \max_{i=1,\ldots,n} x_i' = \mathrm{poly}(||(\mathbf{x}, \varepsilon)||),
$$

and therefore our algorithm runs in time $\mathrm{poly}(n, ||(\mathbf{x}, \varepsilon)||, w_{\max})$. Thus, we have efficiently implemented the separation oracle for our linear program, and the proof is complete. □

Observe that by solving the linear program given in the proof of Theorem 4.13 we explicitly construct a payoff vector in the least core of G. Moreover, we can use the separation oracle described in that proof to check whether a given payoff vector is in the least core. Finally, a straightforward modification of this construction allows us to check if the ε-core is non-empty or to find a payoff vector in the ε-core for a fixed value of ε: all we have to do is to turn our linear program into a linear feasibility program that treats ε as a part of the input. Thus, when weights are given in unary, all least core-related problems admit efficient algorithms.

4.3.1.1 Approximation Scheme for the Least Core

We will now show that the pseudopolynomial algorithm of the previous section can be converted into an approximation scheme. More precisely, we will describe an algorithm that, given a game $G = [N; \mathbf{w}; q]$ and a $\delta > 0$, outputs ε' such that if ε is the value of the least core of G then $\varepsilon \leq \varepsilon' \leq \varepsilon + 2\delta$. The running time of our algorithm is polynomial in the size of the input as well as $1/\delta$, i.e., it is a fully polynomial additive approximation scheme. Subsequently, we will show that it can be transformed into a fully polynomial multiplicative approximation scheme (FPTAS), i.e., an algorithm that outputs ε' satisfying $\varepsilon \leq \varepsilon' \leq (1 + \delta)\varepsilon$ [15, p.111].

Theorem 4.14   There exists an algorithm that, given a game $G = [N; \mathbf{w}; q]$ and a parameter δ, outputs an ε' that satisfies $\varepsilon \leq \varepsilon' \leq \varepsilon + 2\delta$, where ε is the value of the least core of G, and runs in time $\text{poly}(n, \log w_{\max}, 1/\delta)$.

Proof. We will first break the linear program (4.2) into a family of linear feasibility programs $\mathcal{F} = \{\text{LP}_1, \ldots, \text{LP}_t\}$, where $t = \lceil 1/\delta \rceil$. The k-th linear feasibility program in $\mathcal{F}$ is obtained from (4.2) by replacing each appearance of $1 - \varepsilon$ with $k\delta$ (and omitting the objective function). Now, suppose we could solve each of the programs in $\mathcal{F}$, and let k^* be the maximum value of k for which LP_k has a feasible solution. Then if ε is the optimal solution to (4.2), we have $k^*\delta \leq 1 - \varepsilon \leq (k^* + 1)\delta$.

Now, these linear feasibility programs may have exponentially many constraints, so, to solve them, we need a polynomial-time separation oracle. Unfortunately, we do not know how to construct such an oracle. Instead, we will construct a "faulty" separation oracle $\mathcal{O}_k$ for each program LP_k, $k = 1, \ldots, t$ in our family. This oracle has the following property: given a candidate solution $\mathbf{x}$ to LP_k, it either works correctly (i.e., says whether $\mathbf{x}$ is a feasible solution for LP_k, and, if not, outputs a violated constraint), or fails. However, when it fails, it outputs a consolation prize: a feasible solution for LP_{k-1}.

How useful is it to have a faulty separation oracle? Suppose we run the ellipsoid algorithm for each LP_k, $k = 1, \ldots, t$, using $\mathcal{O}_k$ instead of a proper separation oracle. If $\mathcal{O}_k$ fails, we abandon LP_k, but store the solution to LP_{k-1} produced by $\mathcal{O}_k$. If $\mathcal{O}_k$ does not fail, the ellipsoid algorithm outputs the correct answer for this value of k; if it discovers that LP_k has a feasible solution, we store this solution as well.

Let k' be the largest value of k for which our procedure finds a feasible solution for LP_k (coming either from the ellipsoid algorithm or from the separation oracle). Clearly, we have $k' \leq k^*$.

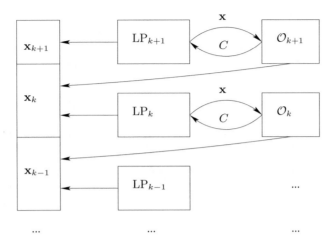

Figure 4.3: Proof of Theorem 4.14: the solver for LP_k uses $\mathcal{O}_k$ as an oracle; a feasible solution $\mathbf{x}_k$ for LP_k may be generated by the solver for LP_k or by $\mathcal{O}_{k+1}$.

On the other hand, when we consider $k = k^*$, one of the following two events can happen: either $\mathcal{O}_k$ works correctly and the ellipsoid algorithm runs to completion producing a feasible solution for LP_{k^*} (which is known to exist by our choice of k^*), or $\mathcal{O}_{k^*}$ produces a feasible solution for LP_{k^*-1}. In either case, among our stored solutions there will be a feasible solution that guarantees a payoff of $(k^* - 1)\delta$ to each winning coalition, and therefore $k' \geq k^* - 1$.

Thus, we have $k^* - 1 \leq k' \leq k^*$, and hence $k'\delta \leq 1 - \varepsilon \leq (k' + 2)\delta$. This means that the algorithm that outputs $1 - k'\delta$ provides an additive 2δ-approximation to the value of the least core.

To prove the bound on the running time, observe that there are $\lceil 1/\delta \rceil$ linear feasibility programs in our list; therefore, it suffices to show that each of them can be processed in time $poly(n, \log w_{max}, 1/\delta)$, or, more specifically, that each of our faulty separation oracles can be implemented within this time bound. We will now show that this is indeed the case.

The subroutine $\mathcal{O}_k$ tries to decide whether there exists a winning coalition that is paid less than $k\delta$ under the given payoff vector $\mathbf{x}$. It starts by setting $\delta' = \delta/n$. It then rounds down each entry of $\mathbf{x}$ to the nearest multiple of δ', i.e., it sets $x'_i = \max_{\ell \in \mathbb{Z}}\{\ell\delta' \mid \ell\delta' \leq x_i\}$ for $i = 1, \ldots, n$. The purpose of this step is to ensure that there are only polynomially many (in n and $1/\delta$) possible payoffs to coalitions, so that one can apply dynamic programming. Observe that we have $0 \leq x_i - x'_i \leq \delta'$ for each $i \in N$. Since each coalition has at most n members, this implies that $0 \leq x(C) - x'(C) \leq n\delta' = \delta$ for each $C \subseteq N$. Now, let

$$z[\ell, j] = \max\{w(C) \mid C \subseteq \{1, \ldots, j\}, x'(C) = \ell\delta'\},$$

where $j = 1, \ldots, n$, $\ell = 0, \ldots, (k-1)n - 1$, and we assume that $\max \emptyset = -\infty$. The values $z[\ell, j]$ are easy to compute by dynamic programming. Let $U = \max\{z[\ell, n] \mid \ell = 0, \ldots, (k-1)n - 1\}$; this is the maximum weight of a coalition whose total payoff under $\mathbf{x}'$ is at most $(k-1)\delta - \delta'$.

Suppose first that $U \geq q$. Then there exists a winning coalition C with $x'(C) \leq (k-1)\delta - \delta'$, and therefore

$$x(C) \leq x'(C) + \delta = k\delta - \delta' < k\delta.$$

Hence, C corresponds to a violated constraint of LP_k. Moreover, we can find C using the standard dynamic programming techniques. Therefore, in this case we have successfully identified a violated constraint, i.e., implemented a separation oracle for LP_k.

Now, suppose that $U < q$, i.e., none of the coalitions that are paid at most $(k-1)\delta - \delta'$ is winning. Since all entries of the modified payoff vector $\mathbf{x}'$ are multiples of δ', this means that each winning coalition is paid at least $(k-1)\delta$ under $\mathbf{x}'$. Consequently, $\mathbf{x}'$ is a feasible solution to LP_{k-1}. Therefore, in this case we allow $\mathcal{O}_k$ to fail and output $\mathbf{x}'$.

The running time of $\mathcal{O}_k$ is polynomial in $\log w_{\max}$, n, and k. Since $k \leq \lceil 1/\delta \rceil$, this proves the desired bound on the running time of $\mathcal{O}_k$. □

To convert our algorithm into an FPTAS, we need the following lemma.

Lemma 4.15   *If the core of a weighted voting game $G = [N; \mathbf{w}; q]$ is empty, i.e., the value of the least core of G is greater than 0, then this value is at least $1/n$.*

Proof. Consider any outcome $\mathbf{x}$ in the least core of G. There exists an agent i such that $x_i \geq 1/n$. As the core of G is empty, it cannot be the case that i is present in all winning coalitions, i.e., there exists a coalition C such that $i \notin C$, $w(C) \geq q$. On the other hand, we have $x(C) \leq 1 - 1/n$, so the value of the least core of G is at least $1/n$. □

We are now ready to prove the main result of this section.

Theorem 4.16   There exists a fully polynomial time approximation scheme for the value of the least core of a weighted voting game, i.e., an algorithm that, given a game $G = [N; \mathbf{w}; q]$ and a parameter δ, outputs an ε' that satisfies $\varepsilon \leq \varepsilon' \leq (1 + \delta)\varepsilon$, where ε is the value of the least core of G, and runs in time $\mathrm{poly}(n, \log w_{\max}, 1/\delta)$.

Proof. Set $\delta' = \frac{\delta}{2n}$ and run the algorithm described in the proof of Theorem 4.14 on G, δ'. Clearly, the running time of this algorithm is polynomial in n, $\log w_{\max}$, and $1/\delta$. Moreover, it outputs ε' that satisfies $\varepsilon \leq \varepsilon' \leq \varepsilon + 2\delta'$. We have $\varepsilon + 2\delta' = \varepsilon + \delta/n \leq \varepsilon(1 + \delta)$, and therefore $\varepsilon' \leq \varepsilon(1 + \delta)$. □

We conclude this section by a brief discussion of the cost of stability and the nucleolus.

From the computational perspective, the cost of stability is very similar to the least core: it is computed by a linear program that has essentially the same structure. Indeed, Bachrach *et al.* [27] show that, while computing the cost of stability in a weighted voting game is NP-hard, this problem admits a pseudopolynomial algorithm as well as an FPTAS.

For the nucleolus, the situation is quite similar. Specifically, Elkind *et al.* [106] prove that the nucleolus is hard to compute. In a subsequent paper, Elkind and Pasechnik [109] show that the nucleolus can be computed in pseudopolynomial time; however, their algorithm is significantly more complicated than the pseudopolynomial algorithms for the least core and the cost of stability.

4.4   VECTOR WEIGHTED VOTING GAMES

We have argued that each weighted voting game is a simple game. One may wonder if the converse is also true, i.e., whether every simple game can be represented as a weighted voting game for a suitable choice of weights and quota. It turns out that this is not the case.

Example 4.17   Consider a game $G = (N, v)$ with $N = \{1, 2, 3, 4\}$ given by

$$v(C) = 0 \qquad \text{if } |C| \leq 1$$
$$v(C) = 1 \qquad \text{if } |C| \geq 3$$
$$v(\{1, 2\}) = v(\{3, 4\}) = v(\{1, 4\}) = v(\{2, 3\}) = 1$$
$$v(\{1, 3\}) = v(\{2, 4\}) = 0$$

We claim that there is no weight vector $\mathbf{w}$ and quota q such that $w(C) \geq q$ if and only if $v(C) = 1$. Indeed, suppose that this condition holds for some $\mathbf{w}$ and q. Then we have

$$w_1 + w_2 \geq q, \quad w_3 + w_4 \geq q,$$
$$w_1 + w_3 < q, \quad w_2 + w_4 < q.$$

The first pair of equations implies $w(N) \geq 2q$, while the second pair implies $w(N) < 2q$, a contradiction.

However, G can be represented as an *intersection* of two weighted voting games, in the following sense. Let $\mathbf{w}^1 = (1, 0, 1, 0)$, $\mathbf{w}^2 = (0, 1, 0, 1)$, $q^1 = q^2 = 1$, and set $G^1 = [N; \mathbf{w}^1; q^1]$, $G^2 = [N; \mathbf{w}^2; q^2]$. Observe that a coalition C is winning in both G^1 and G^2 if and only if it contains both an even-numbered player and an odd-numbered player, i.e., if and only if $v(C) = 1$.

The game considered in Example 4.17 has a counterintuitive property: when two winning coalitions $C_1 = \{1, 2\}$ and $C_2 = \{3, 4\}$ trade members (2 moves from C_1 to C_2 and 3 moves from C_2 to C_1), both of the resulting coalitions are losing. Games where this cannot happen are called *trade robust* (see [251] for the precise definition and a discussion). It can be shown that a game is trade robust if and only if it is a weighted voting game [251].

The construction presented in Example 4.17 can be used to describe other simple games in the language of weighted voting games. That is, we can take $k \geq 1$ weighted voting games $G^1, \ldots, G^k$ over the same set of players N, and define a new simple game $G = (N, v)$ by setting $v(C) = 1$ if and only if C is a winning coalition in each of the underlying games $G^1, \ldots, G^k$. It turns out that any simple game can be obtained in this manner.

Theorem 4.18   For any simple game $G = (N, v)$, there exists a list of weighted voting games $G^1, \ldots, G^k$, where $G^j = [N; \mathbf{w}^j; q^j]$ for $j = 1, \ldots, k$ such that for any coalition $C \subseteq N$ it holds that $v(C) = 1$ if and only if $w^j(C) \geq q^j$ for each $j = 1, \ldots, k$.

Proof. Let $C^1, \ldots, C^k$ be the list of losing coalitions in G. We define the j-th weighted voting game G^j by setting $w_i^j = 0$ if $i \in C^j$ and $w_i^j = 1$ if $i \notin C^j$ (here w_i^j denotes the weight of the i-th player in G^j) and $q^j = 1$. That is, a coalition C is winning in G^j if and only if it contains an agent $i \notin C^j$, or, equivalently, if and only if it is not a subset of C^j.

Consider a coalition C with $v(C) = 1$. By monotonicity, C is not a subset of any losing coalition, so we have $w^j(C) \geq q^j$ for any $j = 1, \ldots, k$. On the other hand, if $v(C) = 0$, then $C = C^j$ for some $j = 1, \ldots, k$, so $w^j(C) = 0 < q^j$. Thus, the theorem is proved.   □

Games that are represented as intersections of k weighted voting games are known as k-weighted voting games or vector weighted voting games. More formally, a *k-weighted voting game*, is given by a set of players N, $|N| = n$, for each player $i \in N$, a k-dimensional weight vector $\mathbf{w}_i = (w_i^1, \ldots, w_i^k)$ whose entries are non-negative integers, and k non-negative integer quotas $q^1, \ldots, q^k$; we write $G = [N; \mathbf{w}_1, \ldots, \mathbf{w}_n; q^1, \ldots, q^k]$. A coalition $C \subseteq N$ is deemed to be winning in G if and only if $\sum_{i \in C} w_i^j \geq q^j$ for all $j = 1, \ldots, k$.

Observe that G can be associated with k weighted voting games $G^1, \ldots, G^k$, where $G^j = [N; (w_1^j, \ldots, w_n^j); q^j]$; these games are called the *component games* of G, and the weight vector of the j-th game is denoted by $\mathbf{w}^j$. Clearly, any vector voting game is a simple game, and Theorem 4.18 shows that the converse is also true.

It is important to note that vector weighted voting games are not theoretical constructs: they feature quite prominently in our lives. For example, the following political systems can be understood as vector weighted voting games [46, 250, 251]:

- The United States Federal system is a 2-weighted voting game, in which the components correspond to the two chambers (the House of Representatives and the Senate). The players are the President, Vice President, senators, and representatives. In the game corresponding to the House of Representatives, senators have zero weight, while in the game corresponding to the Senate, representatives have zero weight. The president is the only player to have non-zero weight in *both* games.

- The voting system of the European Union is a 3-weighted voting game [46]. Specifically, in the Council of the European Union, a law requires the support of 50% of the countries, 62%

G^1 : 29 29 29 29 27 27 14 13 12 12 12 12 12 10 10 10   7   7   7 7 7 4 4 4 4 4 3   255
G^2 :   1   1   1   1   1   1   1   1   1   1   1   1   1   1   1   1   1   1   1 1 1 1 1 1 1 1 1   14
G^3 :170 123 122 120 82 80 47 33 22 21 21 21 21 18 17 17 11 11 11 8 8 5 4 3 2 1 1   620

Figure 4.4: Voting in the Council of the European Union is a 3-weighted voting game.

of the population of the European Union, and 74% of the "commissioners" of the EU. Each member state is a player, so (as of 2011) the players are:

Germany, United Kingdom, France, Italy, Spain, Poland, Romania, The Nether-lands, Greece, Czech Republic, Belgium, Hungary, Portugal, Sweden, Bulgaria, Austria, Slovak Republic, Denmark, Finland, Ireland, Lithuania, Latvia, Slovenia, Estonia, Cyprus, Luxembourg, Malta.

The three component games in the EU voting system are shown in Figure 4.4 (we omit the player set $N = \{1, \ldots, 27\}$, as well as all brackets and parentheses, from the notation). Weights in the first game are assigned according to the number of commissioners the respective member state has. The second game is a simple majority game: every member state gets one vote, and a law must have the support of at least 14 members states. In the third game, weights are assigned in proportion to the population of the respective member state.

Theorem 4.18 allows us to represent any simple game G as a vector weighted voting game; however, the number k of the component games can be quite large (and, in particular, exponential in the number of players n). The number of the component games in a minimal such representation can be interpreted as the inherent complexity of the game. Therefore, given a simple game G, we may be interested in finding the smallest value of k such that G can be represented as a k-weighted voting game. This value of k is called the *dimension* of G; we write $\dim(G) = k$. We emphasize that even if we are given a representation of G as a k-weighted voting game, this does not mean that $\dim(G) = k$: indeed, there may exist a $k' < k$ such that G can be represented as a k'-weighted voting game, so the only conclusion that we can derive is that $\dim(G) \leq k$.

It turns out that there exist simple games of exponentially large dimension. More precisely, for any odd value of n there exists an n-player simple game G such that $\dim(G) \geq 2^{n/2-1}$ (see [251] for a proof).

4.4.1 COMPUTING THE DIMENSION OF A SIMPLE GAME

Our discussion of dimensionality suggests a natural algorithmic question: given a simple game G, can we efficiently determine its dimension $\dim(G)$? Note, however, that if we start with the explicit representation of G, which lists each coalition together with its value, then we cannot hope for an algorithm whose running time is polynomial in n. One way to avoid this issue is to assume that G

is given as a k-weighted voting game, and our goal is to decide whether $\dim(G) = k$. Formally, we define the VWVG-DIMENSION problem as follows:

Name VWVG-DIMENSION.

Instance A vector weighted voting game $G = [N; \mathbf{w}_1, \ldots, \mathbf{w}_n; q^1, \ldots, q^k]$.

Question Is $\dim(G) = k$?

Deineko and Woeginger [91] and, independently, Elkind *et al.* [108] prove that VWVG-DIMENSION is NP-hard, by giving a reduction from a variant of the PARTITION problem. Their hardness result holds even for $k = 2$. However, it is currently not known if VWVG-DIMENSION remains NP-hard when the weights are polynomially bounded; even the case of 0–1 weights (i.e., $w_i^j \in \{0, 1\}$ for all $i = 1, \ldots, n, j = 1, \ldots, k$) is open. Moreover, VWVG-DIMENSION is known to be in Π_2^p (the argument uses Theorem 4.2), but is not known to be in NP or coNP.

An important special case of VWVG-DIMENSION is to decide whether a given vector weighted voting game is equivalent to a weighted voting game. The hardness results of [91, 108] imply that this problem is NP-hard. However, it can be solved in pseudopolynomial time as long as the input game is represented as an intersection of a constant number of weighted voting games [108].

Theorem 4.19 Given a k-weighted voting game $G = [N; \mathbf{w}_1, \ldots, \mathbf{w}_n; q^1, \ldots, q^k]$, where $k = O(1)$, we can decide in time $\mathrm{poly}(n, \max\{w_i^j \mid i = 1, \ldots, n, j = 1, \ldots, k\})$ whether there exists a weighted voting game $G' = [N; \mathbf{w}'; q']$ such that G and G' are equivalent.

Proof. Set $w_{\max} = \max\{w_i^j \mid i = 1, \ldots, n, j = 1, \ldots, k\}$. We construct a linear program $\mathcal{L}$ over variables $w_1', \ldots, w_n', q'$ as follows. For each $C \subseteq N$, we have a constraint that specifies whether $\sum_{i \in C} w_i' \geq q'$ or $\sum_{i \in C} w_i' < q'$, depending on whether C is a winning coalition in G. Additionally, we require that $w_i' \geq 0$ for all $i = 1, \ldots, n$. It is easy to see that $\mathcal{L}$ has a feasible solution if and only if G is equivalent to some weighted voting game.

While this linear program has exponential size, we will now show that it has a separation oracle that can be implemented in time polynomial in $(w_{\max})^k$ and n. Indeed, suppose that we are given a candidate solution $w_1', \ldots, w_n', q'$ (where all w_i' and q' are rational, but not necessarily integer). For each vector of weights $(W_1, \ldots, W_k) \in [0, w_{\max}]^k$ and each $i = 1, \ldots, n$, let

$$X[W_1, \ldots, W_k, i] = \max\{w'(C) \mid C \subseteq \{1, \ldots, i\} \text{ and } \sum_{\ell \in C} w_\ell^j = W_j \text{ for all } j = 1, \ldots, k\}$$

$$Y[W_1, \ldots, W_k, i] = \min\{w'(C) \mid C \subseteq \{1, \ldots, i\} \text{ and } \sum_{\ell \in C} w_\ell^j = W_j \text{ for all } j = 1, \ldots, k\};$$

we use the standard convention that $\max \emptyset = -\infty$ and $\min \emptyset = +\infty$. These quantities can be easily computed by dynamic programming. Specifically, we have

$$X[W_1, \ldots, W_k, 1] = \begin{cases} w_1' & \text{if } w_1^j = W_j \text{ for all } j = 1, \ldots, k \\ -\infty & \text{otherwise} \end{cases}$$

and

$$Y[W_1, \ldots, W_k, 1] = \begin{cases} w_1' & \text{if } w_1^j = W_j \text{ for all } j = 1, \ldots, k \\ +\infty & \text{otherwise.} \end{cases}$$

Moreover, we have

$$X[W_1, \ldots, W_k, i] = \max\{X[W_1, \ldots, W_k, i-1], w_i' + X[W_1 - w_i^1, \ldots, W_k - w_i^k, i-1]\},$$
$$Y[W_1, \ldots, W_k, i] = \min\{Y[W_1, \ldots, W_k, i-1], w_i' + Y[W_1 - w_i^1, \ldots, W_k - w_i^k, i-1]\}.$$

Having computed $X[W_1, \ldots, W_k, n]$, $Y[W_1, \ldots, W_k, n]$ for all $(W^1, \ldots, W^k) \in [0, w_{\max}]^k$, we can easily check if there is a winning coalition in G whose weight under $\mathbf{w}'$ is strictly less than q', or a losing coalition in G whose weight under $\mathbf{w}'$ is at least q'. Thus, we have successfully implemented the separation oracle. Furthermore, it is clear that our implementation runs within the stated time bound; in particular, note that, since $k = O(1)$, running time of the form $\mathrm{poly}(n, (w_{\max})^k)$ is considered to be pseudopolynomial. Hence, $\mathcal{L}$ can be solved in pseudopolynomial time using the ellipsoid method. $\square$

A closely related problem is deciding whether one of the weighted voting games in the representation of G is extraneous. This problem is formalized as follows:

Name VWVG-MINIMALITY.

Instance A k-weighted voting game $G = [N; \mathbf{w}_1, \ldots, \mathbf{w}_n; q^1, \ldots, q^k]$.

Question Is it the case that for each $j \in \{1, \ldots, k\}$ G is not equivalent to the game

$$G^{-j} = [N; \mathbf{w}_1^{-j}, \ldots, \mathbf{w}_n^{-j}; q^1, \ldots, q^{j-1}, q^{j+1}, q^k],$$

given by $\mathbf{w}_i^{-j} = (w_i^1, \ldots, w_i^{j-i}, w_i^{j+1}, \ldots, w_i^k)$ for each $i = 1, \ldots, n$?

Interestingly, Bilbao *et al.* [46] show that the EU voting system is, in fact, *not* minimal: specifically, one can eliminate either the first or third component.

In contrast to VWVG-DIMENSION, VWVG-MINIMALITY can be easily seen to be in NP: to show that a given instance G of this problem is a "yes"-instance, it suffices to exhibit for each $j = 1, \ldots, n$ a coalition C such that C is a losing coalition in G, but a winning coalition in G^{-j} (clearly, the opposite situation is impossible: any winning coalition in G is also a winning coalition in G^{-j}). On the other hand, the hardness proofs for VWVG-Dimension given by [91, 108] directly show that VWVG-MINIMALITY is NP-hard. Moreover, VWVG-MINIMALITY remains NP-hard even for 0–1 weights [108]. The proof of this fact proceeds by reduction from the classic NP-hard problem VERTEX COVER[122], defined below.

Name VERTEX COVER.

Instance An undirected graph $\mathcal{G} = (V, E)$ and a non-negative integer s.

Question Does $\mathcal{G}$ have a vertex cover of size at most s, i.e., is there a vertex set $V' \subseteq V$ with $|V'| \leq s$ such that for any $(u, v) \in E$ we have $\{u, v\} \cap V' \neq \emptyset$?

Theorem 4.20   VWVG-MINIMALITY is NP-hard even if all weights are in $\{0, 1\}$.

Proof. Our proof proceeds by a reduction from VERTEX COVER. Let $(\mathcal{G}, s)$ be an instance of VERTEX COVER, where $\mathcal{G} = (V, E)$, $V = \{v_1, \ldots, v_r\}$, $E = \{e_1, \ldots, e_t\}$. We can assume without loss of generality that $s \leq r - 3$.

We construct an r-player $(t + 1)$-weighted voting game $G = [N; \mathbf{w}_1, \ldots, \mathbf{w}_r; q^1, \ldots, q^t]$ as follows. For $i = 1, \ldots, r$ and $j = 1, \ldots, t$ we set $w_i^j = 1$ if v_i is adjacent to e_j and $w_i^j = 0$ otherwise, and let $q^j = 1$. Also, we set $w_i^{t+1} = 1$ for all $i = 1, \ldots, r$, and let $q^{t+1} = s + 1$.

In this game, coalitions correspond to subsets of vertices. To win the j-th weighted voting game, $j = 1, \ldots, t$, a coalition must cover the j-th edge. Thus, any coalition that wins each of the first t games is a vertex cover for $\mathcal{G}$. Finally, to win the last weighted voting game, the coalition must have at least $s + 1$ members.

Now, if $\mathcal{G}$ does not have a vertex cover of size at most s, the last condition is clearly extraneous: any coalition that wins the first t games also wins the last game. Therefore, a "no"-instance of VERTEX COVER, corresponds to a "no"-instance of our problem. Conversely, suppose that we started with a "yes"-instance of VERTEX COVER. Then it is easy to see that $G^{-(t+1)}$ is not equivalent to G: a coalition that corresponds to a vertex cover of size s or less wins in the former, but not in the latter. Now, consider the game G^{-j} for some $j \in \{1, \ldots, t\}$, and suppose that $e_j = (v_i, v_{i'})$. Then there exists a coalition that wins in G^{-j}, but not in G, namely, $N \setminus \{i, i'\}$. Indeed, this coalition corresponds to a set of vertices that covers all edges except e_j; note that it wins the $(t + 1)$-st weighted voting game since we have assumed $r - 2 \geq s + 1$. Therefore, none of the games G^{-j}, $j = 1, \ldots, t + 1$, is equivalent to G, i.e., we have obtained a "yes"-instance of VWVG-MINIMALITY.   □

The proof of Theorem 4.20 requires k to be large; in particular, it does not go through if k is bounded by a constant. Indeed, Elkind *et al.* [108] show that if both the weights are polynomially bounded and the input game is given as an intersection of a constant number of weighted voting games, VWVG-MINIMALITY can be solved efficiently by dynamic programming.

4.4.1.1 Boolean Combinations of Weighted Voting Games

Vector weighted voting games can be interpreted as conjunctions of weighted voting games. One can also combine weighted voting games according to more complex Boolean formulas: for instance, to win in the game $(G^1 \vee G^2) \wedge G^3$, a coalition must win in one of the games G^1 and G^2 as well as in G^3. This representation for simple games is studied in detail by Faliszewski *et al.* [114]; in particular, Faliszewski *et al.* show that it can be considerably more compact than the representation via vector weighted voting games.

CHAPTER 5

Beyond Characteristic Function Games

So far in this book, we have focused on games in which the payoff to each coalition is completely determined by its composition (i.e., the identities of its members) and can be divided arbitrarily among the coalition members. In this chapter, we investigate what happens if these conditions do not hold. First, we consider games where the payoffs from collaboration are allocated to individual members of the coalition, and cannot be transferred from one player to another; such games are known as *non-transferable utility games*. We then briefly discuss games where payoffs are transferable, but the amount that each coalition can earn is dependent on the coalition structure it belongs to; such games are known as *partition function games*.

5.1 NON-TRANSFERABLE UTILITY GAMES

Recall that characteristic function games are defined as pairs (N, v), with player set N and characteristic function $v : 2^N \rightarrow \mathbb{R}$. When we introduced these games in chapter 2, we did not discuss the semantics of the characteristic function v in detail, but simply stated that the value $v(C)$ is the amount that the players in C could obtain, should they choose to cooperate by forming a coalition. Central to such games is the idea that, in principle, the value $v(C)$ may be distributed amongst the members of C in any way. Because payoff can be distributed arbitrarily in characteristic function games, we say that they are *transferable utility games* (TU games). TU games, in the form of characteristic function games, represent the best-studied class of cooperative games. However, there are many important cooperative settings in which utility is *not* transferable in this way. For example, consider a senior tenured professor A at university X who cooperates with a junior, non-tenured assistant professor B at university Y to write a paper. Both will obtain some benefit from writing the paper, but the benefit that B obtains may well be much greater than the benefit that A obtains, simply because the value added to B's career is greater than the value added to A's, and the benefits that B obtains (enhanced reputation, scientific credibility, standing in the field) cannot easily be transferred from B to A. Such settings in cooperative game theory are the domain of *non-transferable utility games* (NTU games). Characteristic function games are, for obvious reasons, not appropriate for modelling NTU games. In this section, we will focus on NTU games, and in particular, we survey some key NTU game models that have been proposed in the multiagent systems and algorithmic game theory communities.

5.1.1   FORMAL MODEL

We begin by introducing a basic model of NTU games, a counterpart to the characteristic function game model of chapter 2. The basic idea is that each coalition has available to it a set of *choices*, or *consequences*, drawn from some overall set of choices $\Lambda = \{\lambda, \lambda_1, \ldots\}$; we do not require that Λ is finite. We capture the choices available to a coalition C by an NTU version of a characteristic function: in this section, $v(C)$ will denote the *set of choices available to* C. The intended interpretation is that $\lambda \in v(C)$ means that λ is one of the choices available to C, while $\lambda \notin v(C)$ means that C cannot choose λ. Note that we do not mean that C can bring about *all* the choices in $v(C)$ simultaneously, but rather that they can choose *one* of $v(C)$.

Players are not indifferent about choices: they have *preferences* over choices, which we capture in *preference relations*.

Definition 5.1   A *preference relation* on Λ is a binary relation $\succeq \; \subseteq \Lambda \times \Lambda$, which is required to satisfy the following properties:

(1) *Completeness:* For every $\{\lambda, \lambda'\} \subseteq \Lambda$, we have $\lambda \succeq \lambda'$ or $\lambda' \succeq \lambda$;

(2) *Reflexivity:* For every $\lambda \in \Lambda$, we have $\lambda \succeq \lambda$; and

(3) *Transitivity:* For every $\{\lambda_1, \lambda_2, \lambda_3\} \subseteq \Lambda$, if $\lambda_1 \succeq \lambda_2$ and $\lambda_2 \succeq \lambda_3$ then $\lambda_1 \succeq \lambda_3$.

The intended interpretation of $\lambda \succeq \lambda'$ is that the choice λ is preferred at least as much as choice λ'. We let $\lambda \succ \lambda'$ denote the fact that λ is *strictly* preferred over λ' (i.e., $\lambda \succeq \lambda'$ but not $\lambda' \succeq \lambda$). Also, we write $\lambda \sim \lambda'$ if the agent is *indifferent* between λ and λ', i.e., $\lambda \succeq \lambda'$ and $\lambda' \succeq \lambda$.

Each player $i \in N$ is associated with a preference relation $\succeq_i$. Putting these components together, we obtain NTU games.

Definition 5.2   A *non-transferable utility game* (NTU game) is given by a structure $G = (N, \Lambda, v, \succeq_1, \ldots, \succeq_n)$, where $N = \{1, \ldots, n\}$ is a non-empty set of players, $\Lambda = \{\lambda, \lambda_1, \ldots\}$ is a non-empty set of choices, $v : 2^N \to 2^\Lambda$ is the characteristic function of G, which for every coalition C defines the choices $v(C)$ available to C, and, for each player $i \in N$, $\succeq_i \; \subseteq \Lambda \times \Lambda$ is a preference relation on Λ. We will usually assume that $v(\emptyset) = \emptyset$.

It may not be immediately obvious that NTU games generalize transferable utility characteristic function games presented in chapter 2. Observe, however, that a payoff $v(C)$ earned by a coalition C in a characteristic function game $G = (N, v)$ can be associated with a set of vectors

$$X^{v(C)} = \{\mathbf{x} = (x_i)_{i \in C} \in \mathbb{R}^{|C|} \mid \sum_{i \in C} x_i \leq v(C), x_i \geq 0 \text{ for all } i \in C\}.$$

Each vector in $X^{v(C)}$ is a feasible payoff vector for C, or, in other words, a choice that C can enforce. Thus, the characteristic function game $G = (N, v)$ directly corresponds to an NTU game $\widehat{G}$ with the set of players N and the choice space Λ that consists of feasible payoff vectors for coalitions $C \subseteq N$. Each player in N has a natural preference ordering over choices in Λ, preferring the choice under which she receives a higher payoff. This argument shows that any characteristic function game can be viewed as an NTU game, albeit one with an *infinite* set of choices.

5.1.1.1 Outcomes for NTU Games

Recall that an outcome for a characteristic function game $G = (N, v)$ was defined as a pair $(CS, \mathbf{x})$, where CS is a coalition structure over N and $\mathbf{x}$ is a payoff vector, distributing the payoff $v(C)$ obtained by each coalition $C \in CS$ amongst the members of C. We can readily define the NTU analogue of such outcomes.

Definition 5.3   A *choice vector* with respect to a coalition structure $CS = \{C^1, \dots, C^k\}$ of NTU game $G = (N, \Lambda, v, \succeq_1, \dots, \succeq_n)$ is a vector $\mathbf{c} = (\lambda_1, \dots, \lambda_k) \in \Lambda^k$ that satisfies $\lambda_i \in v(C^i)$ for all $i = 1, \dots, k$.

The requirement that $\lambda_i \in v(C^i)$ can be seen as a direct NTU counterpart of the feasibility condition for outcomes $(CS, \mathbf{x})$ in characteristic function games, i.e., that for all $C \in CS$, we must have $\sum_{i \in C} x_i \leq v(C)$.

Definition 5.4   An *outcome for an NTU game* $G = (N, \Lambda, v, \succeq_1, \dots, \succeq_n)$ is a pair $(CS, \mathbf{c})$, where CS is a coalition structure over N and $\mathbf{c}$ is a choice vector for CS. Where $(CS, \mathbf{c})$ is an outcome for an NTU game, and $i \in N$, we let $\mathbf{c}_i$ denote the choice corresponding to the coalition of which i is a member.

It is not difficult to verify that there is a one-to-one correspondence between outcomes of a characteristic function game G and those of its NTU counterpart $\hat{G}$; we leave the proof of this fact as an exercise to the reader.

Subclasses of NTU games

In chapter 2, we identified some important subclasses of characteristic function games. For NTU games, the main subclasses of interest are as follows.

Definition 5.5   An NTU characteristic function v is said to be

(1) *monotone* if $v(C) \subseteq v(D)$ for any two coalitions $C, D \subseteq N$ such that $C \subseteq D$;

(2) *superadditive* if it satisfies $v(C) \cup v(D) \subseteq v(C \cup D)$ for any pair of disjoint coalitions $C, D \subseteq N$.

Depending on the specific domain, the characteristic function may have other useful properties. One possible interpretation for NTU characteristic functions is that $v(C)$ is the set of choices that can be *enforced* by the coalition C, *irrespective of the choices of the players* $N \setminus C$. Where characteristic functions have this interpretation, then they are called *effectivity functions* (see chapter 7).

5.1.1.2   Solution concepts for NTU games

Some solution concepts that we introduced in the context of characteristic function games have a very natural (arguably *more* natural) interpretation in NTU games. We will focus on the core. We formulate the core for NTU games in terms of *objections*.

Definition 5.6   Let $G = (N, \Lambda, v, \succeq_1, \ldots, \succeq_n)$ be an NTU game and let $(CS, \mathbf{c})$ be an outcome of G. Then we say there is an *objection* to $(CS, \mathbf{c})$ if there exists a choice λ and a coalition $C \subseteq N$ such that $\lambda \in v(C)$ and $\lambda \succ_i \mathbf{c}_i$ for all $i \in C$. The *core* of an NTU game G is the set of outcomes for G for which there is no objection.

Example 5.7   Suppose Alice (A) and Bob (B) want a holiday. Alice has \$600 and Bob has \$500. A holiday in Liverpool for one or two people (λ_L) costs \$100, a holiday in Greece for one or two (λ_G) costs \$600, and finally a holiday in Singapore for one or two (λ_S) costs \$1100. Then we have $v(\{A\}) = \{\lambda_L, \lambda_G\}$, $v(\{B\}) = \{\lambda_L\}$, $v(\{A, B\}) = \{\lambda_L, \lambda_G, \lambda_S\}$. The preference relations are as follows:

$$\text{Alice:} \quad \lambda_S \succ_A \lambda_G \succ_A \lambda_L$$
$$\text{Bob:} \quad \lambda_G \succ_B \lambda_S \succ_B \lambda_L$$

Thus, Alice prefers a holiday in Singapore over a holiday in Greece, while Bob prefers Greece over Singapore. For both of them, Liverpool is their least preferred holiday destination. Notice that what Alice and Bob obtain from these various holiday options is personal pleasure, and an associated increase in personal happiness. Such increases in happiness cannot be directly transferred between the two players—the game they are playing really is an NTU game.

   It is easy to see that the core consists of the following outcomes:

$$(\{\{A, B\}\}, (\lambda_S)) \quad (\{\{A, B\}\}, (\lambda_G)).$$

Notice that the characteristic function defined in this example is both monotone and superadditive.

   Solution concepts like the kernel and bargaining set have a similarly straightforward interpretation in NTU games. However, it is arguably less obvious how solution concepts such as the Shapley value can be interpreted. With some appropriate assumptions on the structure of games, however, it is possible to define Shapley-like solution concepts for NTU games. The Maschler–Owen value [175] is among the best known of these NTU values; Hart [137] presents a survey.

5.1.2   HEDONIC GAMES

Hedonic games [39, 55] are a subclass of NTU games in which the set of choices available to each coalition is a singleton, i.e., once a coalition forms, there is only one course of action that is available to it. This effectively means that in hedonic games players have preferences over coalitions that they can join. Indeed, the term "hedonic" stems from the idea that the players can be thought of as "enjoying the pleasure of each other's company". Thus, *the outcome of a hedonic game is simply a coalition structure*. Given a set of players $N = \{1, \ldots, n\}$ and a player $i \in N$, we let $\mathcal{N}_i$ denote the collection of all subsets of N that contain i, that is, $\mathcal{N}_i = \{C \cup \{i\} \mid C \subseteq N\}$. If CS is a coalition structure and $i \in N$ then we denote by CS_i the coalition in CS of which i is a member.

Definition 5.8   A *hedonic game* is given by a structure $G = (N, \succeq_1, \ldots, \succeq_n)$, where $N = \{1, \ldots, n\}$ is the set of players, and for each player $i \in N$ the relation $\succeq_i \subseteq \mathcal{N}_i \times \mathcal{N}_i$ is a complete, reflexive, and transitive preference relation over the possible coalitions of which i is a member. The intended interpretation is that if $C_1 \succeq_i C_2$, then player i prefers to be in coalition C_1 at least as much as in coalition C_2. We define an indifference relationship $\sim_i$ by setting $C_1 \sim_i C_2$ iff $C_1 \succeq_i C_2$ and $C_2 \succeq_i C_1$.

To simplify our presentation, we will often abuse notation by writing $C_1 \succeq_i C_2$ for *arbitrary* coalitions C_1 and C_2, understanding that this is an abbreviation for $C_1 \cup \{i\} \succeq_i C_2 \cup \{i\}$.

5.1.2.1   Solution concepts for hedonic games

We can now define a number of solution concepts for hedonic games [55, p.207–208].

Definition 5.9   Let CS be a coalition structure for hedonic game $G = (N, \succeq_1, \ldots, \succeq_n)$. Then:

- A coalition $C \subseteq N$ *blocks* CS if $C \succ_i CS_i$ for all $i \in C$. The *core* of a hedonic game is the set of coalition structures that are not blocked by any coalition. A hedonic game is *core stable* if it has a non-empty core. Thus if a game has a non-empty core, then there exists a coalition structure such that no set of players in the game would prefer to defect from the coalition structure and form a coalition of their own.

- CS is *individually rational* if $CS_i \succeq_i \{i\}$ for all $i \in N$. If CS is individually rational, then every player does at least as well in CS as it would do alone.

- CS is *Nash stable* if for all $i \in N$ we have $CS_i \succeq_i C_k \cup \{i\}$ for all $C_k \in CS \cup \{\emptyset\}$. Thus Nash stability means that no player would want to join any other coalition in CS (or form a singleton coalition), assuming the other coalitions did not change.

- CS is *individually stable* if there do not exist $i \in N$ and $C \in CS \cup \{\emptyset\}$ such that $C \cup \{i\} \succ_i CS_i$ and $C \cup \{i\} \succeq_j C$ for all $j \in C$. Individual stability means no player could move to another coalition that it preferred without making some member of the coalition it joined unhappy.

- CS is *contractually individually stable* (CIS) if there do not exist $i \in N$ and $C \in CS \cup \{\emptyset\}$ such that:

$$C \cup \{i\} \succ_i CS_i \text{ and } C \cup \{i\} \succeq_j C \text{ for all } j \in C; \text{ and}$$
$$CS_i \setminus \{i\} \succeq_j CS_i \text{ for all } j \in CS_i \setminus \{i\}.$$

A CIS coalition structure is one in which no player can move to another coalition that it prefers so that the move is acceptable to both coalitions it joins and leaves.

These solution concepts are interrelated [55, p.208]:

Theorem 5.10

(1) Individual stability implies contractual individual stability.

(2) Nash stability implies individual stability.

(3) Core stability does not imply Nash stability, nor does Nash stability imply core stability.

(4) Core stability does not imply individual stability.

The reader can verify that the definition of the core for hedonic games is consistent with the definition of the core for general NTU games.

Example 5.11 (From [36].) Consider a hedonic game with $N = \{1, 2, 3\}$ and preference relations defined as follows:

$$
\begin{aligned}
1: &\quad \{1, 3\} \succ_1 \{1, 2, 3\} \succ_1 \{1\} \succ_1 \{1, 2\} \\
2: &\quad \{1, 2\} \succ_2 \{1, 2, 3\} \succ_2 \{2\} \succ_2 \{2, 3\} \\
3: &\quad \{1, 3\} \succ_3 \{1, 2, 3\} \succ_3 \{3\} \succ_3 \{2, 3\}
\end{aligned}
$$

The core of this game is non-empty, and contains just one coalition structure:

$$\{\{1, 3\}, \{2\}\}.$$

This coalition structure is not Nash stable, however, because player 2 wishes to join the coalition $\{1, 3\}$ (i.e., $\{1, 2, 3\} \succ_2 \{2\}$). This coalition structure is individually stable (and hence contractually individually stable): in particular, players 1 and 3 would be worse off if player 2 joined them. The following coalition structure is Nash stable:

$$\{\{1, 2, 3\}\}.$$

To see this, simply observe that if it were not, then a player could do better by unilaterally deviating from the grand coalition and forming a singleton coalition. But for all $i \in \{1, 2, 3\}$, we have $\{1, 2, 3\} \succ_i \{i\}$.

5.1.2.2 Representations for hedonic games

A familiar issue arises: the naive representation of hedonic games (explicitly listing preference orderings $\succeq_i$ for each player i) will be exponential in the number of players. A number of compact representation schemes for hedonic games have been proposed in the literature. We here present a survey of some of the key ones.

Anonymous preferences. A very simple class of preferences in hedonic games is that of *anonymous preferences*: each player's preferences solely depend on the sizes of the coalitions, but not on individual players that appear in these coalitions. In other words, each player i is endowed with a preference relation $\succeq_i'$ over $1, \ldots, n$, and his preference relation over coalitions $\succeq_i$ is given by $C_1 \succeq_i C_2$ iff $|C_1| \succeq_i' |C_2|$.

Individually rational coalition lists (IRCLs). This formalism is based on the idea of eliminating redundant information from the naive representation, i.e., eliminating information that can manifestly play no part in the strategic reasoning of players [36]. Specifically, in IRCL, instead of listing the complete preference ordering $\succeq_i$, we only list the ordering for those coalitions that are preferred by i over the singleton coalition $\{i\}$, that is, the individually rational coalitions. It is easy to see that only such individually rational coalitions can form part of a coalition structure satisfying the solution concepts we defined above for hedonic games. So, for the IRCL representation we define $\succeq_i$ by explicitly listing individually rational coalitions for i, in order, most preferred first, and indicating whether two consecutive coalitions in this order are equally preferred. Formally, the preference list of the player i is represented as $C_1 *^1 C_2 *^2 \cdots *^{r-1} C_r$ where $*^j \in \{\succ_i, \sim_i\}$, $C_j \in \mathcal{N}_i$ for $j = 1, \ldots, r$ and $C_r = \{i\}$.

    Although this representation can eliminate much redundant information, in some cases, IRCLs will be no more succinct than the naive representation. For example, if the worst outcome for a player is working alone (which would happen, for example, if the player needed help to achieve his goal), then IRCLs reduce to the naive representation. Ballester [36] investigated the complexity of the representation, and showed, for example, that checking whether the core of a game is non-empty is NP-complete for the IRCL representation.

Additively separable games. A hedonic game $G = (N, \succeq_1, \ldots, \succeq_n)$ is said to be *additively separable* if there exists an $|N| \times |N|$ matrix of reals M (the *value matrix*) such that

$$C_1 \succeq_i C_2 \quad \text{iff} \quad \sum_{j \in C_1} M[i, j] \geq \sum_{k \in C_2} M[i, k].$$

Thus $M[i, j]$ represents the value of player j to player i. An additively separable game is called *symmetric* if $M[i, j] = M[j, i]$ for all $i, j = 1, \ldots, n$. Symmetric additively separable hedonic games can be viewed as NTU analogues of induced subgraph games discussed in section 3.1.1. If a game is additively separable, then the value matrix M clearly provides a very succinct representation for the game. Of course, not all games are additively separable, and so we cannot use this representation for all games. Sung and Dimitrov [244, 245, 246], Gairing and Savani [120, 121] and Aziz *et al.* [20, 21]

investigated the complexity of this representation. For example, Sung and Dimitrov [244] demonstrated that checking whether a coalition structure is in the core for additively separable preferences is coNP-complete. The tradeoffs between stability and social welfare in additively separable games were studied by Brânzei and Larson [62, 63].

Friends and enemies.   Dimitrov *et al.* [99] introduce two simple subclasses of additive preferences based on the notions of *friends* and *enemies*. Under such preferences, each player views all other players as either friends or enemies; this relationship need not be symmetric. Formally, we have

$$N \setminus \{i\} = F_i \cup E_i \qquad F_i \cap E_i = \emptyset.$$

Now:

- Under *friends appreciation* preferences, players rank coalitions based on how many of their friends they contain, breaking ties based on the number of enemies. Formally, $C_j \succ_i C_k$ iff $|C_j \cap F_i| > |C_k \cap F_i|$ or $|C_j \cap F_i| = |C_k \cap F_i|$ and $|C_j \cap E_i| < |C_k \cap E_i|$. For such preferences, the value matrix is given by $M[i, j] = n$ if $j \in F_i$ and $M[i, j] = -1$ if $j \in E_i$.

- Under *enemies aversion* preferences, the relative importance of friends and enemies is reversed: we have $C_j \succ_i C_k$ iff $|C_j \cap E_i| < |C_k \cap E_i|$ or $|C_j \cap E_i| = |C_k \cap E_i|$ and $|C_j \cap F_i| > |C_k \cap F_i|$. For such preferences, the value matrix is given by $M[i, j] = -n$ if $j \in E_i$ and $M[i, j] = 1$ if $j \in F_i$.

B- and W-Preferences.   Another class of preferences is based on ranking individual players, and ordering coalitions according to the ranks of their worst/best members. In more detail, under $\mathcal{W}$-preferences [65], each player i has a (reflexive, transitive and complete) preference relation $\succeq'_i$ over the set of all players, and he prefers a coalition C_1 to a coalition C_2 if and only if he prefers the worst member of C_1 (according to $\succeq'_i$) to the worst member of C_2.

$\mathcal{B}$-preferences [64] are defined in a similar manner: again, each player i has a preference relation $\succeq'_i$ over individual players, but now we have $C_1 \succeq_i C_2$ if and only if i prefers the best player in C_1 (according to $\succeq'_i$) to the best player in C_2; in addition, ties are broken in favor of smaller sets (otherwise, the grand coalition is the best outcome for everybody).

Hedonic coalition nets.   We conclude with a very general representation scheme called *hedonic coalition nets* [110], which was inspired by Ieong and Shoham's MC-nets representation for characteristic function games [143] (see chapter 3). With hedonic coalition nets, the idea is to define the preferences orders $\succeq_i$ that players have over possible coalitions via a set of rules, with a Boolean formula as the condition of the rule, and a real number on the right hand side of the rule.

To formally define hedonic coalition nets, we assume, as with MC-nets, that we use Boolean logic with variables corresponding to players N. Formally, a *rule for player $i \in N$* is a pair (φ, x), where φ is a Boolean formula over variables N and $x \in \mathbb{R}$. As with MC-nets, we write $C \models \varphi$ to mean that the condition φ applies to coalition $C \subseteq N$ (i.e., that the formula φ evaluates to true under the valuation where every variable appearing in C is set to true and every variable not appearing in C is set to false).

We write a rule (φ, x) for player $i \in N$ using the notation $\varphi \longrightarrow_i x$, omitting the index i where it is clear from the context. Let $\mathcal{R}_i$ be the set of possible rules for player i. A *hedonic coalition net* is a structure $(N, R_1, \ldots, R_n)$ where $N = \{1, \ldots, n\}$ is a set of players, and $R_i \subseteq \mathcal{R}_i$ is a set of rules for player i, for each $i \in N$. The *utility* of a coalition $C \in \mathcal{N}_i$ for a player i is denoted by $u_i(C)$ and is defined:

$$u_i(C) = \sum_{\substack{\varphi \longrightarrow_i x \in R_i : \\ C \models \varphi}} x.$$

The correspondence between utility functions $u_i(\cdot)$ and preference relations $\succeq_i$ is then:

$$\underbrace{C_1 \succeq_i C_2}_{\text{game}} \text{ iff } \underbrace{u_i(C_1) \geq u_i(C_2)}_{\text{net}}.$$

Thus, every hedonic coalition net $(N, R_1, \ldots, R_n)$ defines a hedonic game. Now, it is easy to see that hedonic coalition nets are a *complete* representation for hedonic games, in that any preference relation $\succeq_i$ can be defined using a hedonic coalition net. Moreover, it is frequently a *compact* representation: it is straightforward to construct hedonic coalition nets for which the equivalent naive representation is exponentially larger.

In general, checking whether the core of a hedonic game is non-empty is Σ_2^p-complete for the hedonic net representation, while checking whether a coalition structure is in the core is coNP-complete [110].

5.1.3   QUALITATIVE GAMES

Recall that a simple coalitional game is one in which every coalition $C \subseteq N$ is either winning ($v(C) = 1$) or losing ($v(C) = 0$). In this section, we discuss a class of games, called *qualitative* games, in which it is not *coalitions* that are judged to be either winning or losing, but *individual players*. To put it another way, every player has a goal to achieve, and an outcome *satisfies* a player if that outcome *achieves that player's goal*; otherwise, it is *unsatisfied*.

5.1.3.1   Qualitative coalitional games

Qualitative Coalitional Games (QCGs) were introduced to model this kind of situation [263]. A group of players will cooperate in a QCG to achieve a set of goals that is *mutually satisfying*, i.e., will satisfy all members of the coalition. To model such scenarios, we assume there is some overall set of possible goals Θ, and every agent $i \in N$ is associated with some set of goals $\Theta_i \subseteq \Theta$. Player i wants one of the goals Θ_i to be achieved, but is indifferent about which one. To model cooperative action, we assume that every coalition C has a *set of choices*, $v(C)$, associated with it, representing the different ways that the coalition C could choose to cooperate. Each choice is a set of goals. Formally, the characteristic function for a QCG has the signature:

$$v : 2^N \to 2^{2^\Theta}$$

Notice that if $\Theta' \in v(C)$, then in choosing Θ', coalition C would be choosing to bring about *all* the goals in Θ' simultaneously; this interpretation is subtly different to the interpretation given to NTU characteristic functions. QCGs are formally defined as follows.

Definition 5.12 A *Qualitative Coalitional Game* (QCG), G, is given by a structure $G = (N, \Theta, \Theta_1, \ldots, \Theta_n, v)$, where $N = \{1, \ldots, n\}$ is the set of players, $\Theta = \{\theta_1, \ldots, \theta_m\}$ is a set of *possible goals*, $\Theta_i \subseteq \Theta$ is a set of goals for each player $i \in N$, the intended interpretation being that any of Θ_i would satisfy i, and $v : 2^N \to 2^{2^\Theta}$ is a *characteristic function*, which for every coalition $C \subseteq N$ determines a set $v(C)$ of *choices*, the intended interpretation being that if $\Theta' \in v(C)$, then one of the choices available to coalition C is to bring about *all* the goals in Θ' simultaneously.

Suppose a set of goals $\Theta' \subseteq \Theta$ is achieved. Then Θ' will be said to *satisfy* an agent i if $\Theta_i \cap \Theta' \neq \emptyset$, that is, if agent i gets at least one of its goals achieved. (Notice that an agent has no preferences over goals—it simply wants at least one to be achieved.) A goal set Θ' will be said to be *feasible* for coalition C if Θ' is one of the choices available to C, that is, if $\Theta' \in v(C)$. Finally, a coalition is *successful* if C can cooperate in such a way that all their members are satisfied: more formally, C are successful if there is some feasible goal set Θ' for C such that Θ' satisfies every member of C.

QCGs seem a natural framework within which to study cooperation in goal-oriented systems, but when considering their computational aspects, the usual problem arises: how to represent the characteristic function v? Wooldridge and Dunne [263] proposed a representation based on Boolean logic; this representation is complete (in that every characteristic function can be represented using this scheme), but not always guaranteed to be succinct. The idea is perhaps best illustrated by example. Suppose we have system in which goal θ_1 can be achieved by a coalition C iff C contains agent 1 and either 2 or 3 or both. Then we can represent this characteristic function via the following formula of Boolean logic:

$$\theta_1 \leftrightarrow (1 \wedge (2 \vee 3)).$$

A bit more formally, to represent characteristic functions v, we define a propositional formula, in which Boolean variables in the formula correspond to agents and goals. The idea is that $\Theta' \in v(C)$ iff the formula φ, supposedly representing v, evaluates to true under the valuation that makes the Boolean variables corresponding to Θ' and C true, and all other variables false.

Wooldridge and Dunne [263] investigated the complexity of a range of decision problems for QCGs, using this representation. For example, an interesting possibility in the setting of QCGs is the notion of a *veto player*: we say i is a veto player for j if i is a member of every coalition that can achieve one of j's goals. If i is a veto player for j, then we can say that j is *dependent* on i. Other properties were also defined and discussed.

5.1.3.2 Coalitional resource games

Of course, QCGs have nothing to say about where the characteristic function *comes from*, or how it is *derived* for a given scenario. The framework of *Coalitional Resource Games* (CRGs) gives one answer to this question, based on the simple idea that to achieve a goal requires the consumption, or expenditure of some resources, and that each agent is *endowed* with a profile of resources [264]. Then, coalitions will form in order to pool resources, to achieve mutually satisfactory set of goals.

Definition 5.13   A *coalitional resource game*, G, is given by a structure $G = (N, \Theta, R, \Theta_1, \ldots, \Theta_n, \mathbf{en}, \mathbf{req})$ where $N = \{a_1, \ldots, a_n\}$ is the set of players, $\Theta = \{\theta_1, \ldots, \theta_m\}$ is a set of *possible goals*, $R = \{r_1, \ldots, r_t\}$ is a set of *resources*, $\Theta_i \subseteq \Theta$ is a set of goals for each agent $i \in N$, $\mathbf{en} : N \times R \to \mathbb{N}$ is an *endowment function*, with the intended interpretation that $\mathbf{en}(i, r)$ is the amount of resource r that player i is endowed with, and $\mathbf{req} : \Theta \times R \to \mathbb{N}$ is a *requirement function*, with the intended interpretation that $\mathbf{req}(\theta, r)$ is the amount of resource r that would be required to achieve goal θ.

It is typically assumed that for any goal $\theta \in \Theta$ there exists a resource $r \in R$ such that $\mathbf{req}(r, \theta) > 0$ (every goal has some non-zero resource requirement), as goals that require no resources can be eliminated from a game without altering its strategic structure.

We extend the endowment function $\mathbf{en}$ to coalitions via the function $en : 2^N \times R \to \mathbb{N}$.

$$en(C, r) = \sum_{i \in C} \mathbf{en}(i, r).$$

Similarly, we extend the $\mathbf{req}$ function to sets of goals via the function $req : 2^\Theta \times R \to \mathbb{N}$.

$$req(\Theta', r) = \sum_{\theta \in \Theta'} \mathbf{req}(\theta, r).$$

A set of goals Θ' *satisfies* player i if $\Theta' \cap \Theta_i \neq \emptyset$; we say that Θ' satisfies coalition $C \subseteq N$ if it satisfies every member of C. A set of goals Θ' is *feasible* for coalition C if that coalition is endowed with sufficient resources to achieve all the goals in Θ':

$$feas(\Theta', C) \text{ iff } \forall r \in R : en(C, r) \geq req(\Theta', r).$$

Notice that as we add players to a coalition, the set of goals that this coalition can achieve grows monotonically. That is,

$$feas(\Theta', C) \text{ implies } feas(\Theta', D) \text{ for all } C, D \subseteq N \text{ such that } C \subseteq D.$$

In other words, CRGs are inherently monotone.

We define a function $sf : 2^N \to 2^\Theta$ to return the set of goal sets that both satisfy and are feasible for a given coalition.

$$sf(C) = \{\Theta' \subseteq \Theta \mid feas(\Theta', C) \text{ and } \Theta' \text{ satisfies } C\}.$$

A coalition C are said to be *successful* if $sf(C) \neq \emptyset$.

Example 5.14   (From [264].)   Consider a coalitional resource game G with $N = \{a_1, a_2, a_3\}$, $R = \{r_1, r_2\}$, $\Theta = \{\theta_1, \theta_2\}$, $\Theta_1 = \{\theta_1\}$, $\Theta_2 = \{\theta_2\}$, and $\Theta_3 = \{\theta_1, \theta_2\}$. The endowment function **en** is defined as follows.

$$\begin{aligned}
\mathbf{en}(a_1, r_1) &= 2 & \mathbf{en}(a_1, r_2) &= 0 \\
\mathbf{en}(a_2, r_1) &= 0 & \mathbf{en}(a_2, r_2) &= 1 \\
\mathbf{en}(a_3, r_1) &= 1 & \mathbf{en}(a_3, r_2) &= 2
\end{aligned}$$

And the requirement function as follows.

$$\begin{aligned}
\mathbf{req}(\theta_1, r_1) &= 3 & \mathbf{req}(\theta_1, r_2) &= 2 \\
\mathbf{req}(\theta_2, r_1) &= 2 & \mathbf{req}(\theta_2, r_2) &= 1
\end{aligned}$$

There are eight possible coalitions in the game.

$$\begin{aligned}
C_0 &= \emptyset & C_1 &= \{a_1\} & C_2 &= \{a_2\} & C_3 &= \{a_3\} \\
C_4 &= \{a_1, a_2\} & C_5 &= \{a_1, a_3\} & C_6 &= \{a_2, a_3\} & C_7 &= \{a_1, a_2, a_3\}
\end{aligned}$$

The endowments for these coalitions are summarized in the following table, together with the feasible goal sets for each coalition, and the goal sets that are both feasible for and satisfy each coalition.

	C_0	C_1	C_2	C_3	C_4	C_5	C_6	C_7
$en(C, r_1)$	0	2	0	1	2	3	1	3
$en(C, r_2)$	0	0	1	2	1	2	3	3
$feas(C)$	$\emptyset$	$\emptyset$	$\emptyset$	$\emptyset$	$\{\{\theta_2\}\}$	$\{\{\theta_1\}, \{\theta_2\}\}$	$\emptyset$	$\{\{\theta_1\}, \{\theta_2\}\}$
$sf(C)$	$\emptyset$	$\emptyset$	$\emptyset$	$\emptyset$	$\emptyset$	$\{\{\theta_1\}\}$	$\emptyset$	$\emptyset$

The complexity of decision problems in CRGs was investigated by Wooldridge and Dunne [264]. For example, they showed that the basic problem of determining whether, given a coalition $C \subseteq N$, we have $sf(C) \neq \emptyset$ is NP-complete. Because CRGs relate to resource usage, Wooldridge and Dunne were also able to investigate a range of decision problems relating to resource consumption. For example, suppose we are given a particular budget of resources, and we are asked whether a pair of coalitions can achieve their goals (i.e., whether both can be simultaneously successful), staying within the stated resource bounds. This kind of question arises, for example, when setting targets for pollution control: can some countries achieve their economic objectives without consuming too many pollution-producing resources? Wooldridge and Dunne showed that checking whether a pair of coalitions were in conflict with respect to such a resource bound was coNP-complete. In a follow-on paper, Dunne *et al.* discussed algorithms and bargaining protocols for CRGs [102].

5.1.3.3 Boolean games

Before leaving qualitative games, we should say a few words about cooperative variations of *Boolean games*. Boolean games were introduced by Harrenstein *et al.* [135, 136] and further developed by Bonzon *et al.* [58]. The basic idea in Boolean games is that each player $i \in N$ is associated with a goal θ_i, represented as a Boolean formula over a set of Boolean variables Φ. Each player $i \in N$ is also associated with a subset Φ_i of Φ, with the idea being that Φ_i represents *the set of variables under the control of player i*. Player i has unique control over these variables: his set of strategies consists of all the possible truth assignments to variables in Φ_i. A Boolean game is played by each player choosing a valuation for the variables under her control. The choices made by each player, when taken together, will collectively define a valuation for the overall set of Boolean variables Φ. Player i then obtains utility 1 from this overall valuation if the goal formula θ_i is satisfied by the collective valuation, and gets utility 0 otherwise. Strategic considerations arise in Boolean games because a player's goal formula may include variables controlled by other players, and so whether player i obtains a utility of 1 will depend not just on the choice that i makes, but on the choices of other players in the game. Cooperative variations of Boolean games were investigated by Dunne *et al.* [103]. The following simple example illustrates the basic framework.

Example 5.15   Suppose $N = \{1, 2\}$ with $\Phi = \{p, q\}$, $\Phi_1 = \{p\}$, $\Phi_2 = \{q\}$, $\theta_1 = \neg(p \leftrightarrow q)$, $\theta_2 = p \leftrightarrow q$. Thus player 1 is satisfied by any outcome that gives p and q different values, while player 2 will be satisfied by any outcome that gives them the same value. This game can be viewed as a version of the well-known (non-cooperative) game of *matching pennies* [200, p.17]. The core of this game is empty. The following table shows, for every possible outcome, which agents would prefer to defect (in the interests of readability, we write 0 for falsity and 1 for truth).

p	q	who wants to defect?
0	0	player 1
0	1	player 2
1	0	player 2
1	1	player 1

Alert readers will note that all the deviations in this table are by singleton coalitions, but of course, deviations in general could involve arbitrary coalitions.

Dunne *et al.* demonstrate that checking non-emptiness of the core for cooperative Boolean games is Σ_2^p-complete, and also investigated possible restrictions on goal formulae that would lead to easier decision problems.

5.2   PARTITION FUNCTION GAMES

We have seen that NTU games can model situations that cannot be captured by characteristic function games. However, even NTU games have an important limitation: the choices available to each coalition are assumed to be independent of the actions chosen by the players outside of this coalition. This is not always the case in real life: the choices that a coalition C can make (or, in transferable utility settings, the payoff that C can earn) may depend on the coalition structure formed by the players in $N \setminus C$. Such scenarios are modeled by *partition function games*, which are the subject of this section. As is standard in cooperative game theory literature, we will only define partition function games for settings with transferable utility; however, this definition can be generalized to non-transferable utility settings. We start by presenting a simple example of a cooperative scenario where the value of a coalition depends on the coalition structure it belongs to.

Example 5.16   Consider n apple-growers who all sell their apples in the same market. A coalition of k growers, $1 \le k \le n$, can grow $f(k)$ tons of apples, where $f(\cdot)$ is an increasing convex function with $f(0) = 0$ (e.g., $f(k) = k^2$); the convexity condition reflects the fact that apple-growing is likely to exhibit economies of scale. The market price for apples is determined by their supply and is given by $p(s) = 2f(n) - s$, where s the total supply; note that the convexity of f implies that the price is always non-negative. The total profit earned by a coalition C of size k is the product of the market price and the quantity it produces. In particular, it is equal to $f(k)(2f(n) - f(k) - f(n-k))$ if all other players form a single coalition and to $f(k)(2f(n) - f(k) - (n-k)f(1))$ if all other players remain in singleton coalitions. The convexity condition implies that $f(n-k) \ge (n-k)f(1)$; thus, the growers in C earn more if the other players do not cooperate.

For instance, if $f(k) = k^2$, $n = 4$ and $|C| = 2$, the profit of C is $4(32 - 4 - 4) = 96$ in the former scenario (full cooperation) and $4(32 - 4 - 2) = 104$ in the latter scenario (no cooperation).

Motivated by this example, we will now present the formal definition of partition function games. In these games, *the value of a coalition depends on the coalition structure that it appears in*. We start by introducing the notion of an *embedded coalition*.

Definition 5.17   An *embedded coalition* over N is a pair of the form (C, CS), where $CS \in \mathcal{CS}_N$ is a coalition structure over N, and $C \in CS$. We denote by E_N the set of all embedded coalitions over N.

Now we can define partition function games: these are games that assign a value to each embedded coalition.

Definition 5.18   A *partition function game* G is given by a pair (N, u), where $N = \{1, \ldots, n\}$ is a finite non-empty set of agents and $u : E_N \to \mathbb{R}$ is a mapping that assigns a real number $u(C, CS)$ to each embedded coalition (C, CS).

Example 5.19   Returning to the apple growing game of Example 5.16, with $n = 4$ and $f(k) = k^2$ we have $u(\{1\}, \{\{1\}, \{2, 3, 4\}\}) = 1 \times (32 - 1 - 9) = 22$ and $u(\{2, 3, 4\}, \{\{1\}, \{2, 3, 4\}\}) = 9 \times (32 - 1 - 9) = 198$.

Clearly, any characteristic function game $G = (N, v)$ can be represented as a partition function game $G' = (N, u)$ with $u(C, CS) = v(C)$ for any embedded coalition $(C, CS) \in E_N$. Thus, characteristic function games form a subclass of partition function games, and Example 5.16 illustrates that this inclusion is proper.

Just as for a characteristic function game, an outcome of a partition function game is a coalition structure together with a payoff vector that distributes the value of each coalition among its members. Moreover, most of the solution concepts considered in section 2.2 can be extended to partition function games. However, for some of them there are several non-equivalent approaches to defining the appropriate extension; this is the case, for instance, for the Shapley value (see Michalak *et al.* [181] for an overview). Generally, partition function games appear to be considerably more difficult to work with than characteristic function games, and their computational aspects have received very little attention in algorithmic game theory/multiagent literature so far (see [181, 182, 183, 216] for some notable exceptions). We remark, however, that partition function games model many important scenarios that cannot be captured by characteristic function games and present a very natural direction for future research.

CHAPTER 6

Coalition Structure Formation

In this chapter, we discuss the coalition formation activities undertaken when agents in a multiagent system come together to form teams. We refer to these activities as *coalition structure formation*, and divide them broadly into two classes: *coalition structure generation* activities, undertaken when agents are not selfish but willingly agree to implement the agenda of a single system designer; and *coalition formation activities by selfish rational agents*, where agents choose to participate in coalitions in order to maximize their own utility. We will now discuss both of these models, starting with the former.

6.1 COALITION STRUCTURE GENERATION

Throughout this book, we have considered agents that want to maximize their own utility. However, if the entire system is "owned" by a single designer, then the performance of individual agents is perhaps less important than the *social welfare* of the system, that is, the sum of the values of all coalitions. Of course, optimizing social welfare is straightforward if the underlying game is superadditive (we can assume the grand coalition forms), but, as discussed in chapter 2, many important environments are not superadditive. Therefore, the problem of finding the best partition of agents into teams in non-superadditive settings, which is traditionally referred to as the *coalition structure generation* problem, has been extensively studied. We now state this problem more formally.

A coalition structure CS is said to be *socially optimal* for a characteristic function game G if CS belongs to the set $\arg\max_{CS \in \mathcal{CS}_N} v(CS)$; recall that $v(CS)$ denotes the social welfare $\sum_{C \in CS} v(C)$ of a coalition structure CS. We are interested in finding any element of the set $\arg\max_{CS \in \mathcal{CS}_N} v(CS)$. For presentation purposes, it will be convenient to assume that this set is in fact a singleton—i.e., that there is a *unique* socially optimal coalition structure for G, which we will denote by $CS^*(G)$; however, all of the results discussed in this chapter are independent of this assumption. We will also assume that the value of each coalition is non-negative.

Most of the literature on coalition structure generation assumes that the game is represented by the list of all coalitions together with their values or by an oracle (see section 3.3), and in the rest of the section, we will assume the oracle representation. However, some of the algorithms we present (most notably, that of Sandholm *et al.* [224]) work in a more demanding model where we have an oracle for the values of coalition structures (rather than individual coalitions). Also, we remark that efficient algorithms for coalition structure generation have been recently developed for specific succinct representations (such as MC-nets [196], synergy coalition groups [196], or skill games [30]), or games with a constant number of agent types [22].

Observe that we cannot hope to find $CS^*(G)$ by direct enumeration of all coalition structures: the size of $\mathcal{CS}_N$ is exponential in the number of agents n, and, more importantly, it is huge even relative to the number of coalitions 2^n. The size of $\mathcal{CS}_N$ is the number of partitions of a set of size n, which is known as a *Bell number* B_n [41]. This number can be shown to satisfy $(n/4)^{n/2} \leq B_n < n^n$ (for the lower bound, consider coalition structures where all coalitions have size 2, for n even); see De Bruijn [88] for more precise bounds. This implies that the direct enumeration approach will take super-polynomial time even with respect to the naive representation of the game G, which lists every coalition together with its value. So, can we do better?

6.1.1   DYNAMIC PROGRAMMING

The *dynamic programming* approach to optimal coalition structure generation [220, 267] is perhaps best explained in terms of the notion of a superadditive cover, defined in chapter 2. Recall that the superadditive cover of a game G is the game $G^* = (N, v^*)$ in which the value of each coalition C is the social welfare of the best coalition structure that the agents in C can form. Thus, finding the value of the socially optimal coalition structure in G effectively means computing $v^*(N)$. We will now explain how to compute $v^*(C)$ for *all* coalitions $C \subseteq N$, starting from coalitions of size 1 and 2, and culminating in computing $v^*(N)$. The optimal coalition structure itself can then be recovered using standard dynamic programming techniques. We will need the following simple lemma.

Lemma 6.1   *For any $C \subseteq N$ we have*

$$v^*(C) = \max\{\max\{v^*(C') + v^*(C'') \mid C' \cup C'' = C, C' \cap C'' = \emptyset, C', C'' \neq \emptyset\}, v(C)\}. \quad (6.1)$$

Proof. We will first show that the left-hand side of (6.1) is at least as large as its right-hand side. Clearly, $v^*(C) \geq v(C)$. Now, fix any $C', C'' \subseteq N$ such that $C' \cup C'' = C$, $C' \cap C'' = \emptyset$ and $C', C'' \neq \emptyset$. Let CS' and CS'' be the coalition structures over C' and C'', respectively, such that $v^*(C') = v(CS')$, $v^*(C'') = v(CS'')$. Then $CS' \cup CS''$ is a coalition structure over C with $v(CS' \cup CS'') = v(CS') + v(CS'')$, so we have $v^*(C) \geq v(CS' \cup CS'') = v^*(C') + v^*(C'')$.

We will now argue that the right-hand side of (6.1) is at least as large as its left-hand side. To see this, let CS be a coalition structure over C such that $v^*(C) = v(CS)$. If $CS = \{C\}$, then $v^*(C) = v(C)$ and we are done. Otherwise, let C' be some coalition in CS and let $C'' = C \setminus C'$ (note that $C', C'' \neq \emptyset$). Since $CS \setminus \{C'\}$ is a coalition structure over C'', we have $v^*(C'') \geq v(CS \setminus \{C'\}) = v(CS) - v(C')$. On the other hand, we have $v^*(C') \geq v(C')$. Hence, $v^*(C') + v^*(C'') \geq v(CS) = v^*(C)$. □

Lemma 6.1 tells us that, when computing $v^*(C)$, it suffices to consider partitions of C into two disjoint coalitions (rather than all possible partitions of C, as required by the definition of $v^*(\cdot)$). This allows us to employ the following simple algorithm.

For each $k = 1, \ldots, n$, we compute the values of $v^*(C)$ for all subsets $C \subseteq N$ with $|C| = k$. For $k = 1$, we have $v^*(C) = v(C)$ for all coalitions C of size 1. For an arbitrary value of k, when computing $v^*(C)$ for a coalition C of size k, we consider all partitions of C into two non-empty coalitions C', C'', select the one that maximizes $v^*(C') + v^*(C'')$ (note that $|C'|, |C''| \leq k - 1$, so the values of $v^*(C')$ and $v^*(C'')$ have already been computed), and set $v^*(C) = \max\{v(C), v^*(C') + v^*(C'')\}$.

The algorithm's correctness is guaranteed by Lemma 6.1. Its running time is bounded by $O(\sum_{k=1}^{n} \binom{n}{k} 2^k) = O(3^n)$: for each $k = 1, \ldots, n$, there are $\binom{n}{k}$ coalitions of size k, and for each coalition of size k, there are $2^{k-1} - 1$ ways to split it into two non-empty subcoalitions. Observe that $3^n < (n/4)^{n/2}$ for large enough values of n—i.e., this algorithm avoids searching all coalition structures. Moreover, its running time is polynomial in the number of coalitions 2^n. However, in practice, the dynamic programming algorithm is quite slow. The next section presents an approach that, despite having weaker worst-case performance guarantees, tends to work well in practice.

6.1.2 ANYTIME ALGORITHMS

Sandholm *et al.* [224] developed a technique for finding a coalition structure that is within some *provable bound* from the optimal one. The idea is best explained with reference to an example. Suppose we have a system with four agents, $N = \{1, 2, 3, 4\}$. There are fifteen possible coalition structures for this set of agents; we can usefully visualize these in a *coalition structure graph*—see Figure 6.1. Each node in the graph represents a different possible coalition structure. At level 1 in this graph, we have all the possible coalition structures that contain exactly one coalition: of course, there is just one such coalition structure:

$$\{\{1, 2, 3, 4\}\}.$$

At level 2 in the graph, we have all the possible coalition structures that contain exactly two coalitions—that is, at level 2, we have all possible ways of partitioning the set of agents $\{1, 2, 3, 4\}$ into two disjoint sets. Then, at level 3, we have the possible coalition structures containing 3 coalitions, and so on. (Before proceeding, convince yourself that the graph in Figure 6.1 does indeed capture this information faithfully.) An upward edge in the graph represents the division of a coalition in the lower node into two separate coalitions in the upper node. For example, consider the node with the coalition structure

$$\{\{1\}, \{2, 3, 4\}\}.$$

There is an edge from this node to the one with coalition structure

$$\{\{1\}, \{2\}, \{3, 4\}\},$$

the point being that this latter coalition structure is obtained from the former by dividing the coalition $\{2, 3, 4\}$ into the coalitions $\{2\}$ and $\{3, 4\}$.

The optimal coalition structure $CS^*(G)$ lies somewhere within the coalition structure graph, and so to find it, it seems that we would have to evaluate every node in the graph. But consider the

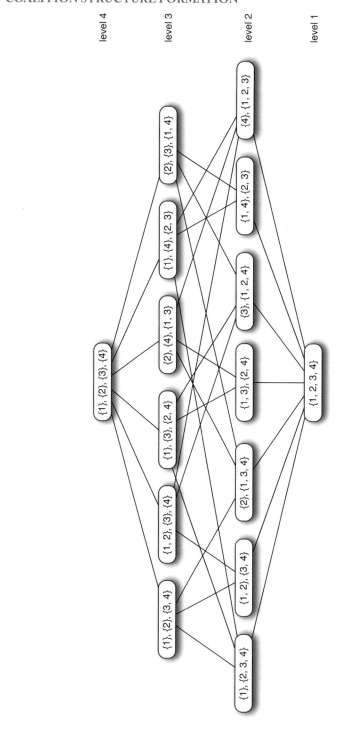

Figure 6.1: The coalition structure graph for $N = \{1, 2, 3, 4\}$. Level 1 has coalition structures containing a single coalition; level 2 has coalition structures containing two coalitions, and so on.

bottom two rows of the graph—levels 1 and 2. Observe that every non-empty coalition appears in these two levels (of course, not every possible *coalition structure* appears in these two levels). Now, suppose we restrict our search for the optimal coalition structure to *just* these two levels—we go no higher in the graph. Let CS' be the best coalition structure that we find in these two levels, and let $CS^*(G)$ be the best coalition structure overall. We will now argue that $v(CS^*) \leq n \cdot v(CS')$—in words, the value of the best coalition structure we find in the first two levels of the graph is within a factor of $\frac{1}{n}$ from the optimum.

To see this, let C^* be a coalition with the highest value of all possible coalitions—i.e., $v(C^*) \geq v(C)$ for all $C \subseteq N$. Since every possible coalition appears in at least one coalition structure in the first two levels of the graph, and the values of all coalitions are non-negative, we have $v(CS') \geq v(C^*)$. On the other hand, CS^* has at most n coalitions, and the value of each coalition in CS^* is at most $v(C^*)$, so we have $v(CS^*) \leq n \cdot v(C^*) \leq n \cdot v(CS')$, which is what we set out to prove. Thus, although searching the first two levels of the graph does not guarantee to give us the *optimal* coalition structure, it *does* guarantee to give us one that is no worse than $\frac{1}{n}$ of the optimal.

This idea immediately gives us a simple algorithm for finding good coalition structures: search the bottom two levels of the coalition structure graph, keeping track of the best coalition structure found so far. Now, suppose we have more time available to continue the search for a good structure beyond the bottom two levels of the graph. How should we implement this search in the best way possible? Sandholm *et al.* [224, p.218] propose the following algorithm:

(1) Search the bottom two levels of the coalition structure graph, keeping track of the best coalition structure seen so far;

(2) Now continue with breadth-first search starting at the *top* of the graph, again keeping track of the best coalition structure seen so far, and continue until either the time is up, or else we have visited all the parts of the graph not considered in stage (1);

(3) Return the coalition structure with the highest value seen.

Experimentally, this approach has been shown to perform quite well. It particular, it has the advantage of being *anytime*—that is, it is guaranteed to produce incrementally better solutions if given more time or computational resources. While this is not a primary requirement, it is highly desirable in many applications.

Note, however, that even searching the first two levels of the coalition structure graph is fairly time-consuming: the second level contains 2^{n-1} coalition structures. Indeed, Sandholm *et al.* also show that to get within a factor of $1/n$ of the optimal solution, we need to search at least 2^{n-1} structures.

These techniques were refined and extended in [84, 218]. Specifically, Dang and Jennings [84] also use the coalition structure graph, but instead of partitioning the graph into levels, they define and search particular graph subsets; after a subset is searched, the solution lies within a certain bound from the optimal. In contrast, Rahwan *et al.* [218] do not employ Sandholm *et al.*'s coalition structure graph, but use a different representation for the subspaces of the search space. In their representation, each subspace is a collection of coalition structures that uniquely corresponds to an integer partition of the number of agents—e.g., $P_{\{1,1,2\}}$, with indices (parts) being the sizes of coalitions in the structure. They then employ branch-and-bound techniques to search each subspace. Overall, their approach works by obtaining bounds on subspaces defined using the aforementioned representation, and then making use of these bounds to select subspaces to search. They demonstrate experimentally that they can reach near-optimal solutions faster than the anytime approaches discussed above.

Interestingly, the dynamic programming algorithm described in section 6.1.1 has a natural interpretation in the context of the coalition structure graph: It first evaluates every possible movement in the graph, and then starts moving upwards from the bottom node through a path of connected nodes until an optimal node is reached, after which no movement is beneficial. In fact, it would be able to find the optimal structure even if some edges in the graph were missing, as long as there is at least one path that leads to $CS^*(G)$. This observation is the basis of the *improved dynamic programming (IDP)* algorithm that was proposed by Rahwan and Jennings [215], which is able to find the optimal coalition structure by using substantially less time and memory. Subsequently, Rahwan and Jennings [214] presented a *hybrid* approach employing IDP to prune the search space, and then switching to Rahwan *et al.*'s anytime algorithm [218] to focus on the most promising regions of the space. The hybrid algorithm was experimentally shown to be significantly faster than both of its ingredients. A different way of turning the dynamic programming algorithm into an anytime algorithm is proposed by Service and Adams [230]; their approach produces an approximation scheme, i.e., a family of algorithms whose running time depends on the desired approximation ratio.

An important applications domain for coalition structure generation is disaster management [152, 206], where agents need to form coalitions to help humans in need. For such settings, a *distributed* coalition structure generation method is desirable; one such algorithm was recently put forward by Michalak *et al.* [183]. Also, Banerjee and Kraemer [37] and Rahwan *et al.* [217] built on previous work to provide anytime algorithms for partition function game environments.

6.2   COALITION FORMATION BY SELFISH RATIONAL AGENTS

The coalition structure generation problem implicitly assumes that *benevolent* agents can be put together into teams by a common owner/system designer, who aims to partition the players into coalitions so as to maximize the social welfare. However, in many real-world settings, one has to deal with rational agents forming coalitions in order to serve their own interests. Simply put, agents are not benevolent, but rather selfish; though they recognize the need to work with others, they are interested in getting as much as possible out of some kind of contract they sign with their partners to share the outcomes of their collective efforts. In this section, we will examine coalition formation activities that take into account the fact that agents are self-interested.

6.2.1   COALITION FORMATION VIA BARGAINING

Sometimes agents can conduct explicit negotiations as a part of a *bargaining process* to form coalitions. Such *coalitional bargaining procedures* can be seen as (non-cooperative) extensive-form games [174, 191, 200]. A wide variety of such games can be found in the literature [74, 111, 139, 186, 199, 208, 229, 266]; we will now describe several representative examples.

In the model of Okada [199], at each round of bargaining a random proposer chosen out of the set of participating agents puts forward a proposal that consists of a coalition and a payoff vector for this coalition. If the proposal is accepted, the coalition abandons negotiations. After each round, the values of all coalitions go down by a fixed discount factor. Okada essentially characterizes the subgame-perfect equilibria (SPE)[1] of the resulting game. Specifically, he shows that if the SPE are assumed to be *stationary* and the proposer is chosen *uniformly at random* at each round, then there is *no delay of agreement in equilibrium*—i.e., in equilibrium, any proposal made at round one is agreed upon by the interested agents immediately; this result is closely related to Rubinstein's [221] seminal work on the alternating-offers model of two-person bargaining. The stationarity of the SPE means that the agents choose identical equilibrium strategies in subgames involving the same set of active agents—the proposals and responses of the agents in the t-th round of the game depend only on the set of players active at that round, and not on past history. Under the stationarity assumption, the equilibrium proposals and responses of the players are given as solutions to a payoff maximization problem. We remark that the assumption that the proposer is chosen randomly is essential for this result: Chatterjee *et al.* [74] present a discounted multi-person coalitional bargaining model with a *fixed* proposer order, and show that it *can* result in a delay of agreement in equilibrium.

Moldovanu and Winter [186] study bargaining games assuming non-transferable rather than transferable utility. The order of proposers in the game depends on the responders' replies: the first responder to refuse a proposal becomes the next *initiator*, and can propose a coalition and a payoff vector (out of a set of available payoff vectors) to its members—or, it can pass the initiative to another player. When all potential members accept a proposal, the formed coalition abandons the game. There is no discounting of coalitional values over time. Similarly to Okada, Moldovanu and Winter focus on stationary subgame-perfect equilibria (SPE). They show that if a bargaining strategy profile is an *order independent equilibrium (OIE)*—that is, an SPE that remains an equilibrium and leads to the same payoff allocation for any choice of proposer in a sequential coalitional bargaining game—then the resulting payoffs must be in the core; conversely, if the coalition formation game has subgames with non-empty cores, then for each payoff vector there exists an OIE with the same payoff.

The reader may notice that, even though Moldovanu and Winter view the coalition formation process as a non-cooperative game, their results show that the outcome of this process can be linked with a traditional cooperative solution concept, namely, the core. Many other coalitional bargaining procedures considered in the literature converge to outcomes in the core under certain

---

[1]In an extensive form game of perfect information, a subgame-perfect equilibrium is a strategy profile whose restriction to any subgame following any history in the game is a Nash equilibrium of the subgame [174, 200].

assumptions [111, 139, 186, 208, 229, 266]. The line of work that connects outcomes of coalitional bargaining with cooperative solution concepts is known as the *Nash program for cooperative games* (see chapter 7).

6.2.2   DYNAMIC COALITION FORMATION

The coalition formation algorithms mentioned in the previous subsection assume that the potential coalition members employ a sequential bargaining process. This might not always be the case, as coalitions may form and then fall apart in an online fashion: *ad hoc* political or market alliances within a "fluid" environment provide an example. In general, the coalition structure might evolve due to any of several internal or external factors, such as agents' decisions to abandon their current alliances, new agents entering the system, fluctuations in the communication and computational ability of partners, and so on. We will be using the term "dynamic" to refer to such coalition formation paradigms.

An early paper on this topic by Shehory and Kraus [235] focused on a (non-bargaining) three-stage process by which coalitions emerge. At the first stage of the process, agents calculate the values of potential coalitions in a distributed fashion. To achieve this, a protocol determines how agents may approach each other and commit to taking up the responsibility of calculating coalitional values, after acquiring full information about the capabilities of their potential partners. The second stage consists of an iterative greedy process through which, given the calculated coalitional values, the agents decide upon the preferred coalitions and form them. At the third stage of the process, the benefits from forming coalitions are distributed among the agents. As the complexity of this approach is exponential in the number of agents, certain heuristics to reduce computational costs may be employed. One such heuristic (subsequently used in many coalition formation algorithms), is to bound the size of each coalition by a small constant in order to reduce communication and value calculation costs.

Arnold and Schwalbe [8] study coalitional stability under dynamic coalition formation. The coalition formation process they describe can be viewed as a Markov process, and allows for exploration of suboptimal "coalition formation actions". At each stage of the process, a given configuration of a coalition structure and associated allocations of payoffs, or *demands*, is assumed to be in place. Then, with some small probability γ, each player independently decides whether she wants to try to get a higher payoff by moving to a new coalition. If she decides to move, her choice of new partners is described by a "non-cooperative best-reply rule": a player switches coalitions if her expected payoff in the best available coalition exceeds her current payoff, and she demands the most she can get subject to feasibility constraints. Formed coalitions do not abandon the process, and there are no explicit proposers or responders: rather, the process evolves by the agents adjusting their coalitions and demands as long as changes are feasible.

In some more detail, the process where all players adopt the best-reply rule corresponds to a finite Markov chain with state space

$$\Omega = \{\omega = (CS, \mathbf{d}) \mid CS \in \mathcal{CS}, \mathbf{d} \in \times_{i \in N} D_i\},$$

where $\mathcal{CS}$ is the space of all possible coalition structures and D_i is the set of allowable demands of player i, which is restricted to be finite for reasons of computational tractability.

The transition probabilities can be described as follows. In each state ω, each player i flips a biased coin with bias γ to decide whether he should adjust his demand. All players make these choices simultaneously and independently. Then, each adjusting player i computes the maximum payoff that he can get by moving to an existing coalition in ω or forming a singleton coalition:

$$d_i(\omega) = \max_{S \in \mathcal{CS} \cup \{\emptyset\}} v(S \cup \{i\}) - \sum_{\substack{j \neq i \\ j \in S}} d_j.$$

All adjusting players then move simultaneously: each player i attempts to join a coalition that guarantees him a payoff of $d_i(\omega)$, randomizing over all such coalitions if there are several of them. This may result in an infeasible outcome. For instance, suppose that i moves to a coalition S, but at the same time some other agent $j \in S$ leaves S: if $v(\{S \setminus \{j\}\} \cup \{i\}) < v(S \cup \{i\}) - d_j$, i's demand cannot be satisfied. Whenever this is the case, every agent i' in S receives his reservation value $v(\{i'\})$. However, if γ is small, this turn of events is unlikely.

The finite Markov chain described above is shown to have at least one absorbing state. If the players are assumed to *explore* with myopically suboptimal actions, the process can be viewed as "best reply with experimentation". Arnold and Schwalbe prove that if the core is non-empty, each core allocation corresponds to an absorbing state of this new process, and each such state can be associated with a core allocation. Furthermore, if the core is non-empty, the process is proved to converge to a core allocation with certainty.

Konishi and Ray [156] study a different dynamic coalition formation process: Coalitions move to a new state (i.e., to a new coalition structure, accompanied by a corresponding payoff allocation) only if the move is profitable to all members of the coalition. The agents have common beliefs about the probability with which the state transitions may occur.

Augustine *et al.* [10] study the dynamics of coalitional games with submodular characteristic functions under three natural profit-sharing schemes. In their model, agents are partitioned into teams, and the members of each team are paid according to the chosen profit-sharing scheme. The agents may change teams myopically, aiming to increase their payoff. Augustine *et al.* show that this process converges to a Nash equilibrium and provide bounds on the social welfare of the states that can be achieved after a polynomial number of moves.

In the multiagent systems community, an early task-oriented approach to the dynamic coalition formation problem was proposed by Sycara *et al.* [248, 249], who developed the Reusable Environment for Task Structured Intelligent Network Agents (RETSINA) framework. RETSINA is distributed system, which allows potentially heterogeneous agents to enter and leave the system dynamically, as well as to collaborate and team up with others in order to decompose and execute tasks. This system can deal with heterogeneous agents who can take up different roles in order to interact with users, gather and pass information around, or execute tasks.

6.2.3 COALITION FORMATION UNDER UNCERTAINTY

In all approaches considered so far in this book, the underlying assumption was that the values of all coalitions are known with certainty. However, in most realistic coalition formation settings, this assumption cannot be reasonably expected to hold. Indeed, agents might be uncertain about the capabilities of potential partners, the resources available, the eagerness of partners to make use of their skills or available resources, the deadlines the coalitions might need to meet, the possibility of unexpected obstacles and failures, and so on. In what follows, we give a brief overview of some of the literature that attempted to address those very real issues.

Uncertainty about partners' capabilities and/or resources

In an early paper on this topic, Ketchpel [151] explicitly describes a two-agent auction protocol to facilitate coalition formation in the face of uncertain rewards. Kraus *et al.* [160] look at the infinite horizon alternating offers model of bargaining where agents take the passage of time into account. They examine a case where agents need to negotiate how to share a joint resource, and have incomplete information about each other. In order to decide on accepting or rejecting offers, the agents update beliefs regarding other agents' types; the updates are based on other agents' responses to previous offers. A finite set of agent types is assumed, each type having a different utility function which depends on its resource usage. However, the negotiation in any given period is *bilateral*, since it is assumed that no more than two agents need to share the same resource at any given time period. Specifically, the agents play two different roles: one of them already has access to the resource and is using it during the negotiation process (gaining over time), while the other is waiting to use the resource (losing over time).

Shehory and Kraus [237, 238] develop coalition formation algorithms which account for uncertainty regarding the capabilities of agents. Agents rely on information communicated to them by their potential partners in order to form initial estimates of the partners' capabilities. Payoff allocation is addressed in [238], where two algorithms for forming coalitions of self-interested agents in non-superadditive environments are presented. The first algorithm guarantees an approximation to optimal social welfare *assuming a specific coalitional configuration*, while the second algorithm is negotiation-based and is anytime ε-stable. Both algorithms deal directly with expected payoff allocation, and use kernel-based coalition formation mechanisms.

Shehory *et al.* [239] also assume that agents communicate in order to learn the capabilities of their partners, and develop coalition formation algorithms to achieve agent collaboration in the RETSINA framework. This work focuses on serving the needs of the team (in other words, on maximizing social welfare) and does not deal with payoff allocation issues.

Kraus *et al.* [158, 159] deal with coalition formation in the "Request for Proposal" domain. In this domain, the agents come together to perform tasks comprised of subtasks; each subtask should be performed by a different agent. The agents may not know the costs that other agents incur by performing a subtask, but they *do* know the overall payoff associated with performing a task and the capabilities of the other agents. The tasks are allocated by means of an auction, and agents use heuristics to form teams in order to bid in this auction. In [158], the authors assume that all agents in a coalition divide the gained surplus equally, which does not necessarily encourage the truthful reporting of the costs during the coalition formation process. The subsequent paper [159] presents two alternative payoff allocation strategies, namely (1) distributing payoffs in proportion to the agents' costs and (2) using a kernel-based payoff division scheme. The authors also propose combining these allocation strategies with a *compromise-based* approach, where the agents may be willing to receive less payoff as long as they do get to form the coalition. Their experiments show that compromises facilitate the formation of successful coalitions. However, Kraus, Shehory and Taase do not fully address the issues of strategic behavior: the set of agents' strategies is limited to the heuristics listed in the paper, and deviations to strategies that do not appear in this list are not considered. Also, this work does not tackle iterative coalition formation: once a coalition is formed, it "walks away" from negotiations and cannot be decomposed.

Banerjee and Sen [38] address uncertainty regarding payoffs to members entering a coalition. They do not deal with the process of coalition formation itself or payoff allocation. Rather, they focus on the problem of "coalition selection": an agent has imperfect summary information on the anticipated payoff from joining a coalition, and has to choose one coalition over another after a fixed number of allowed interactions with them. This "summary information" is provided by a multinomial probability distribution over possible payoffs for joining the coalition. The proposed mechanism for choosing a coalition makes use of this distribution, and also employs an arbitration mechanism to resolve ties. When the number of allowed interactions is limited, the proposed mechanism notably outperforms a *maximization of expected utility* mechanism in terms of selecting the most beneficial coalition. In the absence of such limit, however, the former mechanism reduces to the latter.

Blankenburg *et al.* [52] implement a coalition formation process that allows agents to progressively update *trust* values, by communicating their private estimates regarding task costs and coalition valuations. They use cryptographic techniques and develop a payment protocol that incentivizes the agents to truthfully report their valuations. However, their mechanism requires extensive inter-agent communication, and relies on computing optimal coalition structures and kernel-stable solutions (both of which are computationally intensive tasks).

Chalkiadakis *et al.* [70] study coalitional games in which every player has a *type*, the value of each coalition is defined as a function of the coalition members' types, and players have (non-probabilistic) *beliefs* about each other's types. These beliefs need not be consistent: two players may have different beliefs about the third player's type. Chalkiadakis *et al.* give a definition of the core for this setting, and characterize simple games with beliefs that have a non-empty core. However, they do not provide a mechanism that allows the players to update their beliefs according to the performance of the coalitions that they form.

Stochastic cooperative games

A more rigorous handling of coalition value uncertainty is provided by Suijs *et al.* [242, 243], who introduce *stochastic cooperative games (SCGs)*. These games are described by a set of agents, a set of coalitional actions, and a function assigning a random variable with finite expectation to each coalitional action, representing the expected payoff to the coalition when this action is taken.

An important issue in stochastic environments is that the agents are uncertain about coalitional payoffs, and hence about their own payoffs. To deal with this issue, Suijs *et al.* [242, 243] use *relative* payoff shares, i.e., allow the agents to place relative demands on the fractional share of the realized payoff. This directly accounts for the allocation of unexpected gains or losses. Formally, let $\mathbf{d} = (d_1, \ldots, d_n)$ be the payoff demand vector; these demands are assumed to be observable by all agents. For any player i in a coalition C, his *relative demand* is defined to be $r_i = d_i / \sum_{j \in C} d_j$. If coalition C receives a reward R, each player $i \in C$ receives a payoff of $r_i R$.

These papers provide theoretical foundations for games with payoff uncertainty and describe classes of games for which the core of a SCG is non-empty. Suijs *et al.* [243] also discuss the effect of agents' risk attitude on the existence of a core allocation within a specific subclass of SCGs. However, Suijs *et al.* do not explicitly model a coalition formation process. Moreover, their model does not consider the possibility of incomplete information about partners' capabilities, and assumes that agents have *common expectations* regarding expected coalitional values. This assumption is relaxed in the work on Bayesian cooperative games, which we describe next.

6.2.3.1 Bayesian cooperative games

The term "Bayesian cooperative/coalitional games" has been introduced independently by several groups of authors, namely, Myerson [190, 192], Chalkiadakis *et al.* [66, 67, 68, 72], and Ieong and Shoham [144]. We start by presenting the model of Chalkiadakis *et al.*, as it builds directly on the work of Suijs *et al.* that was described in the previous subsection. We discuss the work of Ieong and Shoham towards the end of this section, and postpone the discussion of Myerson's papers to the next section, as it fits well with the mechanism design approaches that are discussed there.

An important feature of Chalkiadakis *et al.*'s approach is that it models coalition value uncertainty via *agent type uncertainty*. In more detail, Chalkiadakis *et al.* study a class of *Bayesian coalition formation problems (BCFPs)*, where each agent i has *private probabilistic beliefs* $B_i(\mathbf{t}_{-i})$ about the private *types* (or, *capabilities*) $\mathbf{t}_{-i}$ of others—and, thus, beliefs $B_i(\mathbf{t}_C)$ about the type profile $\mathbf{t}_C$ of the members of any coalition C. This type uncertainty is then translated into uncertainty about the coalitional values, and is used to inform the agents of their potential gains from the outcomes of a coalition formation process.

As an illustrative example, consider a group of contractors, say plumbers, electricians, and carpenters coming together to collaborate on various construction projects. Each contractor possesses trade-specific skills of various degrees corresponding to their types, e.g., a carpenter might be highly skilled or moderately incompetent. Any group of contractors that joins together to form a coalition will receive a payoff for the house they build. The payoff received for a house depends on the type of project undertaken and its resulting quality, which in turn depends on the quality of each team member and potential synergies or conflicts among them. Agents are uncertain about the types of potential partners, but their beliefs are used to determine a distribution over coalitional outcomes and expected coalitional value. It is these beliefs that influence the coalition formation process and the stability of the resulting coalition structure. Each coalition must also decide which collective action to take. For instance, a team of contractors may have a choice of what type of housing project to undertake (e.g., a high-rise building in Toronto or a townhouse estate in Southampton). The outcome of such coalitional actions is stochastic, but it is influenced by the types of agents in the coalition. This too plays a key role in determining the value of the coalition. Deliberations about team formation are complicated by the fact that uncertainty about partner types influences the choice of coalitional actions (e.g., what type of house to build) as well as payoff division (i.e., how to split the revenue generated).

More formally, each coalition C in BCFPs takes a coalitional action $a \in A_C$ with a *stochastic* outcome $s \in S$ that occurs with probability $\Pr(s \mid a, \mathbf{t}_C)$ and results in a reward $R(s)$. As in the work of Suijs *et al.*, each agent i submits a demand d_i, and the relative demand of any agent i in a coalition C is his share $r_i = d_i / \sum_{j \in C} d_j$ of total payoff of C; in what follows, we will denote the demand vector by $\mathbf{d} = (d_1, \ldots, d_n)$, and the relative demand vector for a coalition C by $\mathbf{r}_C = (r_i)_{i \in C}$. An outcome of a Bayesian coalition formation process is a *configuration* $\langle CS, \mathbf{a}, \mathbf{d} \rangle$, where CS is a coalition structure, $\mathbf{a} = (a_C)_{C \in CS}$ is a vector of coalitional actions, and $\mathbf{d}$ is a demand vector; we will denote by C_i the coalition in CS that contains player i.

In this model, type uncertainty translates into coalitional value uncertainty. Given their privately-held beliefs about the types of potential partners, agents formulate expectations about the value of any:

1. *Coalition–action pair* (C, a)

$$Q_i(C, a) = \sum_{\mathbf{t}_C \in T_C} B_i(\mathbf{t}_C) \sum_s \Pr(s \mid a, \mathbf{t}_C) R(s).$$

   This is i's expectation regarding the value of coalition C taking action a, given i's beliefs about the profile $\mathbf{t}_C$ that describes the types of C's members.

2. *Coalition*

$$V_i(C) = \max_{a \in A_C} Q_i(C, a).$$

   This is the immediate value that i assigns to coalition C, i.e., the value of that coalition executing its optimal action, given i's beliefs.

3. *Coalitional agreement* $\langle C, a, \mathbf{d}_C \rangle$

$$p_i^i(C, a, \mathbf{d}_C) = r_i V_i(C).$$

   This is what i expects to get out of an agreement specifying that: (a) i will be a member of C, (b) C will execute action a, and (c) the payoffs will be divided using the demand vector $\mathbf{d}_C$.

Agent i can also form a belief about the value of an agreement $\langle C, a, \mathbf{d}_C \rangle$ to any agent $k \in C$, by assuming that the latter is given by $p_k^i(C, a, \mathbf{d}_C) = r_k V_i(C)$: this is the value of this agreement given i's beliefs and her lack of knowledge about k's privately held beliefs.

Stability now has to be ensured *given the agents' beliefs*: Ideally, no agent should have an incentive to suggest a change to the current configuration—i.e., no alternative coalition that she could (reasonably) expect to join would offer her a better expected payoff than what she expects to receive given the action choice and allocation agreed upon by the coalition to which she belongs. Failing that, for any agent that may believe that a deviation can be beneficial, the beliefs of her potential partners should preclude such deviation. Thus, stricter or looser forms of stability can be defined, depending on the level of agreement one would require among the beliefs of various agents. To illustrate, we list here three variants of the *Bayesian Core (BC)* for games with coalition structures [66].

The first of these, the *weak BC*, consists of all configurations $\langle CS, \mathbf{a}, \mathbf{d} \rangle$ such that there exists no coalition all of whose members believe that they (personally) can be better off in it (in terms of expected payoffs, given some choice of action) than they are within $\langle CS, \mathbf{a}, \mathbf{d} \rangle$. The agents' beliefs, in every $C \in CS$, "coincide" in the weak sense that there is a payoff allocation $\mathbf{d}_C$ and some coalitional action a_C that is commonly believed to ensure the best payoff.

Definition 6.2 (Weak Bayesian Core)   A configuration $\langle CS, \mathbf{a}, \mathbf{d} \rangle$ (or, equivalently, $\langle CS, \mathbf{a}, \mathbf{r} \rangle$) is in the *weak Bayesian core* of a BCFP if there is no coalition $S \subseteq N$, demand vector $\mathbf{d}_S$ and action $b \in A_S$ such that $p_i^i(S, b, \mathbf{d}_S) > p_i^i(C_i, a_{C_i}, \mathbf{d}_{C_i})$ for all $i \in S$, where $\mathbf{d}_{C_i}$ is the restriction of $\mathbf{d}$ to the coalition C_i.

In BCFPs with *continuous* payoffs, the transferability of utility means that if a new coalition makes any agent strictly better off with respect to his beliefs without making other agents worse off with respect to their own beliefs, then it can make all members strictly better off through a suitable adjustment of relative demands. However, if the set of demands that an agent can submit is finite, this is no longer the case. We can then define a stronger version of the Bayesian core, by demanding that there is *no* agent who believes that there exists a coalitional agreement that can make him strictly better off *and* which other coalition members believe to be not harmful, according to *their* own beliefs.

Definition 6.3 (Strict Bayesian Core)   A configuration $\langle CS, \mathbf{a}, \mathbf{d} \rangle$ is in the *strict Bayesian core* of a BCFP if there is no coalition $S \subseteq N$, demand vector $\mathbf{d}_S$ and action $b \in A_S$ such that $p_j^j(S, b, \mathbf{d}_S) \geq p_j^j(C_j, a_{C_j}, \mathbf{d}_{C_j})$ for all $j \in S$ and $p_i^i(S, b, \mathbf{d}_S) > p_i^i(C_i, a_{C_i}, \mathbf{d}_{C_i})$ for some $i \in S$.

The strict Bayesian core is stricter than the weak Bayesian core in the sense that the former is a *subset* of the latter (for a finite set of permissible demands).

There is a third, distinct core-stability concept that can be defined both for finite and continuous demands; we will call it the strong BC. For a configuration to be in the strong BC, no agent should believe that there is an agreement that can make him better off and that he expects all partners to accept; his expectations about the partners' behavior are based on his own subjective view of their expected payoff. This differs inherently from the weak and the strict core in that the agent assesses *his own* beliefs about the value of an agreement to his partners.

Definition 6.4   (Strong Bayesian Core)   A configuration $\langle CS, \mathbf{a}, \mathbf{d} \rangle$ is in the *strong Bayesian core* if there is no coalition $S \subseteq N$, demand vector $\mathbf{d}_S$ and action $b \in A_S$ such that for some $i \in S$ it holds that $p_i^i(S, b, \mathbf{d}_S) > p_i^i(C_i, a_{C_i}, \mathbf{d}_{C_i})$ and $p_j^i(S, b, \mathbf{d}_S) \geq p_j^i(C_j, a_{C_i}, \mathbf{d}_{C_j})$ for all $j \in S \setminus \{i\}$.

It is easy to see that, if the agents' beliefs coincide, the strict BC and the strong BC coincide.

Building on this Bayesian model, Chalkiadakis *et al.* [66, 67] presented a *dynamic process with random proposers* to be used for coalition formation under type uncertainty, and demonstrated its convergence to the *strong BC*. The process can be seen as an extension of the best reply with experimentation process of Arnold and Schwalbe [8], as it explores potentially profitable structures; however, it does so given *beliefs* and by using explicit proposals. Chalkiadakis *et al.* also proposed a heuristic algorithm for coalitional bargaining under uncertainty regarding the capabilities of potential partners [66, 68]. The algorithm uses iterative coalition formation with belief updating based on the observed actions of others during bargaining, and attempts to simulate the play in a *perfect Bayesian equilibrium (PBE)* [174] of a coalitional bargaining game. The algorithm is shown to perform well empirically, and can be combined with belief updates after observing the results of coalitional actions. Finally, they provided explicit links to multiagent reinforcement learning, employing Bayesian learning techniques to tackle the agents' *sequential decision making problem* in repeated coalition formation environments [66, 69]. We present this learning approach in some more detail in the next section.

Ieong and Shoham [144] also propose a model for Bayesian coalitional games, which, however, instead of representing the uncertainty w.r.t. agent types, describes it w.r.t. possible *worlds* of games in which the agents may participate. That is, agents possess a common prior over the set of possible coalitional games, but also have private information about the true state of the world, captured by their *information partitions*. Specifically, each agent is assumed to partition the possible states of the world into *information sets*; the worlds in each information set are indistinguishable from the agent's point of view. Given this model, Ieong and Shoham proceed to define an *ex-ante*, an *ex-interim* and an *ex-post* core—that is, the set of outcomes that are stable before a world is drawn; after a world is drawn, when each agent is made aware of the information set to which the world belongs in his partition of the worlds; or after the true world is made known, respectively. They also show that checking for core emptiness reduces to a linear feasibility problem when agents are risk-neutral. Their *ex-interim* core is closely linked to the Bayesian core concept of Chalkiadakis *et al.*, but their focus is on the stability of the grand coalition. Several other technical differences between the two concepts exist, with an important one being that Ieong and Shoham's core allows for an infinite level of mutual agent modelling (w.r.t. other agents' expected payoff).

6.3   COALITION FORMATION AND LEARNING

In the presence of uncertainty, agents who participate in *repeated* coalition formation activities can benefit from attempting to figure out the privately-held information determining the actions and performance of potential partners—that is, from trying to *learn* other players' types. Indeed, it is only natural for a rational agent to ponder questions such as: *"How should I choose partners to work with in a coalition if I am not fully aware of their capabilities? What if I have the opportunity to choose partners repeatedly? How do I update my beliefs regarding the value of coalitional agreements? How do I maximize my long-term rewards from such a process?"* We view these as key questions for research on learning in cooperative games.

Learning in (non-cooperative) games has been actively studied by the artificial intelligence and game theory research communities [119]. In particular, techniques derived from the field of *reinforcement learning (RL)* [247] have provided much inspiration when incorporating learning into stochastic and repeated game settings. Broadly speaking, RL agents attempt to learn from experience gathered while acting in some environment. The environment can be modeled as a *Markov Decision Process (MDP)* [213] controlled by the agent, whose goal is to maximize the long-term reward received.

In more detail, an MDP is a 4-tuple $\langle S, A, R, D \rangle$, where S is a (finite) state set, A is a (finite) action set, R is the reward function, and D is the transition dynamics. The transition dynamics D is a family of distributions $\Pr(s, a, s')$, where $\Pr(s, a, s')$ is the probability of reaching state s' after taking action a at state s, and $R(s, a, r)$ is the probability that reward r is obtained when executing action a at state s. The agent's goal is to construct a Markovian policy $\pi : S \rightarrow A$ maximizing the expected sum of future discounted rewards over an infinite horizon. This policy, and its value, $V^*(s)$ at each $s \in S$, can be computed using standard algorithms, such as value or policy iteration [213, 247]. From an RL point of view, an MDP can be viewed as a complete specification of the environment that satisfies the Markov property [247].

In an RL world, it is assumed that either the reward or the transition dynamics, or both, are unknown to the agent. Then, to achieve its goal of long-term reward maximization, a central problem that an RL agent has to tackle is the *exploration-exploitation* one: should he be content with what has been discovered so far, or should more parts of the environment be explored? At least in principle, *Bayesian* methods allow for the solution of this problem: they assume some prior density P over possible dynamics D and reward distributions R (a *"belief state"*), which is updated with experience gained, and allow agents to explore optimally [42, 90, 101]. The use of appropriate *conjugate* families of distributions can help alleviate the problem of the proper Bayesian updating of beliefs [89]. However, in many cases the applicability of Bayesian methods is limited, as the number of reachable belief states grows exponentially with the time horizon.

A multitude of techniques have been proposed to solve the RL problem in a suboptimal but computationally efficient manner, with some of them enjoying considerable success, such as the widely used Q-learning method [260]. Unfortunately, a technique's success in single-agent worlds does not readily translate into success in multiagent worlds, where one has to learn how to align her action choices with those of others, since the effects of an agent's actions are directly influenced by the actions of others. Learning in cooperative games, in particular, can be quite challenging. This is because agents have to agree on coalitional actions and payoff shares at various stages of the learning process, and a single agent does not have complete control over such future decisions. As a result, there has been only a limited amount of work on learning in coalitional games.

One, can, however, build on the theoretical strength of Bayesian techniques, and attempt to develop a Bayesian RL framework for sequentially optimal repeated coalition formation under uncertainty. To begin, observe that agents participating in coalition formation activities will generally face two forms of uncertainty: (1) type uncertainty—i.e., uncertainty regarding the types (or capabilities) of potential partners and (2) uncertainty regarding the results of coalitional actions. Repeated interaction provides agents with an opportunity to tackle both of these types of uncertainty: rational agents might explicitly take actions that reduce specific type or action uncertainty, rather than trying to optimize the myopic value of the "next" coalition they join.

Starting from this realization, Chalkiadakis and Boutilier [66, 67, 68, 69] employ tractable Bayesian RL techniques within a framework where agents come together repeatedly in episodes, during which they can form new coalitions and take coalitional actions. In the construction project example mentioned above, for instance, after one set of housing projects is completed, the agents have an opportunity to regroup, forming new teams. Of course, the outcomes of previous coalitional actions provide each agent with information about the types of his previous partners. In our example, receiving a high price for a house may indicate to a plumber that the electrician and carpenter he partnered with were highly competent. Agents update their beliefs about their partners based on those prior outcomes and use these updated beliefs in their future coalitional deliberations. For instance, an agent may decide to abandon its current partners to join a new group that she believes may be more profitable. This model brings together coalition formation under uncertainty—specifically, the model for Bayesian coalition formation problems (BCFP) discussed earlier in this chapter—with Bayesian reinforcement learning, to enable the agents to take better coalition formation decisions and reach more profitable coalitional agreements.

More formally, the learning process can be described as follows. Consider an infinite horizon model, in which a set of agents N faces a Bayesian coalition formation problem at each stage $0 \leq t < \infty$. The BCFP at each stage is identical, except that, at each stage, each agent may enter the coalition formation process with *updated* beliefs that reflect her past interactions with previous partners: At stage t, each agent i has beliefs B_i^t about the types of all agents (including the certain knowledge of her own type). Coalitions are formed, resulting in a configuration $\langle CS^t, \mathbf{d}^t, \mathbf{a}^t \rangle$, with coalition structure CS^t, demand vector $\mathbf{d}^t$ (and induced relative demand vector $\mathbf{r}^t$), and action vector $\mathbf{a}^t$. Each coalition $C \in CS^t$ takes its agreed upon action a_C^t and observes the *stochastic outcome state* s that is realized. Such outcomes are local; that is, they depend only upon the action a_C taken by C and the type vector $\mathbf{t}_C$. The dynamics of this process is determined by $\Pr(s \mid a, \mathbf{t}_C)$, where s is the stochastic outcome of coalitional action a executed by coalition C when its type profile is $\mathbf{t}_C$. Limited observability is assumed: agent i observes only the outcome of the action taken by its own coalition C_i, not those of other coalitions. Each stochastic outcome s is associated with a specific reward $R(s)$ for the coalition, with each i in C obtaining its relative share $r_i R(s)$.

Once coalitional agreements are reached, actions are executed, and outcomes observed at stage t, the process moves to stage $t + 1$ and repeats. Let γ $(0 \leq \gamma < 1)$ be a discount factor applied to future expected rewards, and let R_i^t be a random variable denoting agent i's realized reward share at stage t of the process. Then, i's goal is to maximize $\sum_{t=0}^{\infty} \gamma^t R_i^t$.

With this process in place, the critical exploration-exploitation tradeoff in RL is embodied in the tension between (1) forming teams with partners whose types are known with a high degree of certainty (e.g., stay in one's current coalition) or (2) forming teams with partners about whom much less in known, in order to learn more about these new partners' abilities. This tradeoff can be made optimally by relying on the concept of *value of information* [141]. Specifically, the agent can model this problem as a partially observable Markov decision process (POMDP) [147]. Its solution, then, determines agent policies that value *coalition formation actions*—i.e., choices for ⟨coalition, coalitional action, payoff allocation⟩ triplets—not just for their immediate gains but also because of the *information* they provide about the types of others and the values of potential coalitions and coalitional agreements [69].

It is reasonably straightforward to formulate the optimality equations for this POMDP; however, certain subtleties arise because of an agent's lack of knowledge of other agent beliefs. In more detail, let agent i have beliefs B_i about the types of other agents. Let $Q_i(C, a, \mathbf{d}_C, B_i)$ denote the *long-term, sequential value* that agent i places on being a member of coalition C with agreed-upon action a and demands $\mathbf{d}_C$, realizing that after this action is taken the coalition formation process will repeat. This is accounted for using Bellman equations [43] as follows:

$$Q_i(C, a, \mathbf{d}_C, B_i) = \sum_s \Pr(s \mid C, a, B_i)[r_i R(s) + \gamma V_i(B_i^{s,a})] \qquad (6.2)$$

$$= \sum_{\mathbf{t}_C} B_i(\mathbf{t}_C) \sum_s \Pr(s \mid a, \mathbf{t}_C)[r_i R(s) + \gamma V_i(B_i^{s,a})]$$

$$V_i(B_i) = \sum_{C: i \in C, \mathbf{d}_C} \Pr(C, a, \mathbf{d}_C \mid B_i) Q_i(C, a, \mathbf{d}_C, B_i), \qquad (6.3)$$

where $B_i^{s,a}$ is agent i's updated belief state obtained after joining coalition C, demanding d_i, and executing coalitional action a_C which resulted in s. Recall that r_i is i's relative demand of the payoff received by i's coalition C; hence, $r_i R(s)$ describes i's reward. $V_i(B_i)$ reflects the value of belief state B_i to i, deriving from the fact that given beliefs B_i, agent i may find itself participating in any of a number of possible coalitional agreements, each of which has some Q-value. Of note is the fact that agent i considers the (expected, discounted) value of being in its updated belief state $B_i^{s,a}$ when computing the value of any coalitional agreement: decisions at future stages may exploit information gleaned from the current interaction.

In contrast to typical Bellman equations, the value function V_i cannot be defined by maximizing Q-values. This is because the choice that dictates reward, namely, the coalition that is formed, is not under complete control of agent i. Instead, i must predict, based on his beliefs, the probability $\Pr(C, a, \mathbf{d}_C \mid B_i)$ with which a specific coalition C (to which it belongs) and a corresponding action-demands pair $\langle a_C, \mathbf{d}_C \rangle$ will arise as a result of negotiation. Nevertheless, with this in hand, the value equations provide the means to determine the long-term value of any coalitional agreement. Solving POMDPs is, however, computationally intractable. Therefore Chalkiadakis and Boutilier develop and evaluate several computational approximations that construct sequential policies more effectively.

The Bayesian approach described above aims for optimality, and attempts to handle all aspects of the coalition formation problem. Other approaches to learning in coalitional games, however, employ certain simplifying assumptions—such as ignoring the payoff allocation part of the coalition formation scenario, and focusing on social welfare maximization under uncertainty instead. These simplifications ensure tractability and may be appropriate for many applications. For instance, Abdallah and Lesser [2] also utilize reinforcement learning in their approach to "organization-based coalition formation". They assume an underlying organization which guides the coalition formation process, and use Q-learning to optimize the decisions of coalition managers, who assess communication or action processing costs. However, agents are assumed not to be selfish, and there is no attempt to solve the payoff allocation problem. Furthermore, managers are assumed to possess full knowledge of the capabilities of their subordinates.

Klusch and Gerber [153] design and implement a simulation-based dynamic coalition formation scheme which can be instantiated using a variety of computational methods and negotiation protocols. Their framework can be employed for the development, implementation and experimental evaluation of different coalition formation and payoff allocation algorithms (even fuzzy or stochastic coalition formation environments). Gerber's thesis [123] explores several different algorithms for learning to form coalitions within the aforementioned scheme.

Finally, departing from the use of RL approaches in cooperative games, Procaccia and Rosenschein [211] employ *Probably Approximately Correct (PAC) learning* [256] to identify the minimum winning coalitions in simple cooperative games. They provide a learning algorithm for such games, and show that restricting the set of minimum coalitions to a single coalition, or a dictator (a player whose presence is both necessary and sufficient for a coalition to be winning), greatly reduces the difficulty of learning this particular class. However, they also show that restricting attention to other interesting subclasses of simple games does not ease learning significantly.

CHAPTER 7

Advanced Topics

Our aim in this chapter is to briefly survey some advanced topics, and some topics at the frontiers of contemporary research. Much of the work discussed in this chapter is very recent, and presents interesting directions for future work. We also give a brief overview of some real-world applications of cooperative game theory. In contrast to the remainder of this book, in this chapter, we eschew technical definitions and formalism, and instead simply aim to provide a high-level summary of the key concepts and problems, together with a collection of key references.

7.1 LINKS BETWEEN COOPERATIVE AND NON-COOPERATIVE GAMES

Having discussed a number of models and solution concepts for cooperative games, we would now like to elaborate on connections between games with and without cooperation. We will argue that, by taking the possibility of cooperation into account, we can obtain a better understanding of some scenarios that are usually modeled as non-cooperative games; conversely, in many cases, cooperative solutions concepts can be justified in non-cooperative terms.

7.1.1 COOPERATION IN NORMAL-FORM GAMES

Aumann [11] and Aumann and Peleg [14] observed that, if players in a normal-form game are allowed to cooperate, the resulting game can be interpreted as an NTU game. Recall that a non-cooperative (normal-form) game is given by a set of players N, and, for each player $i \in N$, a set of actions A_i and a utility function u_i that maps action profiles, i.e., elements of $\prod_{i \in N} A_i$, to real numbers (the Prisoner's Dilemma in chapter 1 was an example of such a game). Now, suppose that we have a game G represented in this form, but we stipulate that agents can make binding commitments—i.e., form coalitions. What is the resulting cooperative game?

This question does not have an obvious answer: the difficulty is that the choices available to a coalition $C \subseteq N$ (i.e., the payoffs that the members of C can ensure for themselves) depend on the actions of players in $N \setminus C$. There are two standard ways of handling this issue: the first is to assume that agents in C have to choose their actions before observing what the agents in $N \setminus C$ do (and are pessimistic about the actions of these players), and the second is to assume that agents in

C get to see what the agents in $N \setminus C$ chose to do before selecting their own action profile. These two models give rise to the notions of α-effectivity and β-effectivity, respectively.

Definition 7.1   Let $G = \langle N, (A_i)_{i \in N}, (u_i)_{i \in N} \rangle$ be a normal-form game. Consider a coalition $C \subseteq N$, and let $\mathcal{A}_C = \prod_{i \in C} A_i$, $\mathcal{A}_{-C} = \prod_{i \notin C} A_i$. Then

- C is *α-effective* for a payoff vector $(x_i)_{i \in C}$ if there exists an action profile $\mathbf{a}_C \in \mathcal{A}_C$ such that for any action profile $\mathbf{a}_{-C} \in \mathcal{A}_{-C}$ it holds that $u_i(\mathbf{a}_C, \mathbf{a}_{-C}) \geq x_i$ for all $i \in C$.

- C is *β-effective* for a payoff vector $(x_i)_{i \in C}$ if for any action profile $\mathbf{a}_{-C} \in \mathcal{A}_{-C}$ there exists an action profile $\mathbf{a}_C \in \mathcal{A}_C$ such that $u_i(\mathbf{a}_C, \mathbf{a}_{-C}) \geq x_i$ for all $i \in C$.

By identifying the choices available to a coalition with payoff vectors that this coalition is γ-effective for (where $\gamma \in \{\alpha, \beta\}$) and defining the agents' preferences over the choices in a natural way, we obtain an NTU game G^γ that is said to be *γ-associated* with the original normal-form game G. We remark that the games G^α and G^β can be very different. If we allow transfers among players, we can use a similar construction to build a characteristic function game γ-associated with G. The notion of effectivity (and its generalizations) plays an important role in the analysis of a wide range of scenarios, most notably, in social choice theory [3, 187].

Observe that a representation formalism that succinctly encodes a normal-form game G can also be used to describe the cooperative version of this game G^γ (for $\gamma \in \{\alpha, \beta\}$); however, computing the characteristic function of G^γ from this representation may be difficult. This issue has been recently investigated by Bachrach *et al.* [31], who tackle this problem in the context of congestion games; however, the issue merits further research.

Reasoning in terms of coalitional behavior may also be helpful when dealing with the problem of *equilibrium selection*, i.e., identifying the most plausible Nash equilibrium in a game that admits multiple equilibria. Indeed, an outcome is arguably more appealing if it is robust not just against deviations by individual agents (as required by the definition of Nash equilibrium), but also against deviations by *groups* of agents. This line of thinking gives rise to the notions of *strong equilibrium* [11] and *coalition-proof equilibrium* [44]. Under both solution concepts, an outcome of a non-cooperative game is deemed to be stable if no coalition can profitably deviate from it; the difference between the two concepts is that the former assumes that the deviators behave myopically, while the latter requires the deviations to be self-enforcing. In both cases, it is assumed that the deviating players cannot make transfers to each other, i.e., a deviation should be beneficial to each member of the deviating group.

Finally, we remark that some real-life scenarios can be plausibly modeled both as cooperative games and as normal-form games. For instance, the minimum spanning tree game (section 3.1.4) has a natural non-cooperative counterpart, where an agent's strategy is described by his contribution to each edge, and the edge is considered to be bought if the total contribution to this edge exceeds its cost. Hoefer [140] builds on the results of Deng *et al.* [96] to analyze the relationship between the core of a cooperative game and the strong equilibrium of its non-cooperative version; his results apply to several classes of cost-sharing games.

7.1.2   NON-COOPERATIVE JUSTIFICATIONS OF COOPERATIVE SOLUTION CONCEPTS

While we have presented a number of solution concepts in this book, we usually did not provide an explicit procedure that allows players to arrive to these solutions. One exception from this rule is section 6.2.1, which discusses bargaining processes that lead to outcomes in the core. In general, such a procedure may take the form of a non-cooperative extensive-form game that produces an outcome that belongs to a certain coalitional solution concept. This procedure can then serve as a non-cooperative justification of this concept, showing that such outcomes can indeed arise as a result of interaction among selfish agents. This research agenda was initiated by Nash [193] and became known as the *"Nash program"* [228].

A number of significant results have been obtained within the Nash program, including for example the work surveyed in section 6.2.1. The successes of the Nash program are not limited to the core: for instance, Hart and Mas-Colell [138] describe a bargaining game that produces the Shapley value payoffs in the subgame-perfect Nash equilibrium, and Serrano [227] shows that for bankruptcy problems there is a negotiation procedure that produces the nucleolus as its unique outcome. For the Bayesian model, Chalkiadakis *et al.* [66, 72] prove that if the Bayesian core of a coalitional game (and of each subgame) is non-empty, then there exists an equilibrium of the corresponding bargaining game that produces a Bayesian core element; conversely, if there exists a coalitional bargaining equilibrium with certain properties, then it induces a Bayesian core configuration. We refer the reader to a survey by Serrano [228] for a more comprehensive overview of the Nash program.

The Nash program focuses on sequential procedures that result in desirable outcomes, i.e., extensive-form games. However, certain cooperative solution concepts also arise as outcomes of normal-form games. For instance, Potters and Tijs [209] show that each cooperative game can be associated with a matrix game so that the nucleolus of the cooperative game coincides with the strategy in the unique proper equilibrium of the corresponding matrix game. Similarly, Aziz *et al.* [23, 24] argue that least core payoffs in a simple coalitional game correspond to maxmin strategies in a related non-cooperative zero-sum game.

7.1.3   PROGRAM EQUILIBRIUM

One interesting perspective on the links between cooperative and non-cooperative games was provided by Tennenholtz in his notion of *program equilibrium* [252]. Program equilibria are appealing for several reasons. First, they provide an answer to a question that we have rather glossed over in this book: namely, *what exactly are binding agreements?* Second, as we will see, they provide a novel perspective on the applications of computing techniques in cooperative game theory.

To understand the idea of program equilibria, let us return to the non-cooperative game that we discussed in chapter 1: the Prisoner's Dilemma. The Prisoner's Dilemma is called a dilemma because the dominant strategy equilibrium of the game (mutual defection, i.e., both players confessing) leads to an outcome that is strictly worse for *both* players than another outcome (mutual cooperation, i.e., keeping quiet). Many researchers have attempted to "recover" cooperation from the Prisoner's Dilemma—to try to find some way of explaining how and why mutual cooperation can rationally occur (see, e.g., [48, 49] for extensive references and critical discussion on this topic). One answer is to *play the game more than once*. It is well-known in the game theory literature that there is an important difference between playing a game once and playing a game a number of times. The set of Nash equilibria of a repeated game will typically include outcomes that are not simply iterations of the Nash equilibria of the component game. Specifically, the Nash Folk Theorems [200, p.143] tell us that every strategy profile leading to players obtaining better than their minimax payoff can be achieved in an infinitely repeated game. The simple technical device used in the proof of Nash's Folk Theorems is a construction known as a *trigger strategy*, which can be understood as "punishing" players that do not "cooperate". Such punishment is possible in repeated games because the players will meet each other again, and a group of players can punish a deviant player by enforcing that player's minimax payoff. Of course, in one-shot games, it seems that trigger strategies are not possible since the players will not meet again in the future. Thus, in the Prisoner's Dilemma, the problem is that each prisoner must commit to either cooperate or defect, whereas intuitively, what each prisoner wants to do is *make his commitment contingent on the commitments of the other player*. The difficulty, however, is making this precise: Tennenholtz's notion of program equilibria does exactly this.

Tennenholtz proposed that players in a game enter complex strategies expressed as *programs that may be conditioned on the programs submitted by others*. In [252], the conditions permitted on programs were restricted to be comparisons of program text; that is, comparing whether the "source files" for two programs were the same. Using this scheme, Tennenholtz proposed that a player in the Prisoner's Dilemma game should enter a program strategy as follows[1]:

```
IF HisProgram == MyProgram THEN
  DO(COOPERATE);
ELSE
  DO(DEFECT);
END-IF.
```

The intended meaning of this program strategy is as follows: `HisProgram` is a string variable containing the text of the other player's program strategy, `MyProgram` is a string variable containing the text of the program in which it is referred, `DO(...)` indicates a commitment to choosing one of the available actions, and "==" denotes the string comparison test. Now, suppose one player enters the program above: *then the other player can do no better than enter the same program*, yielding the overall outcome of mutual cooperation as an equilibrium. Notice that the strategy program given above is, of course, very much like a trigger strategy in iterated games, and using this basic construct, Tennenholtz was able to prove a version of Nash's Folk Theorem for one-shot games.

[1]We will not here formally define the "program strategy language" used by Tennenholtz—the key features should be easy to understand from the example.

7.2 USING MECHANISM DESIGN FOR COALITION FORMATION

When agent types—and hence, coalitional values—are not known with certainty, agents may act strategically by misreporting their types. Thus, one may need to use the tools of *mechanism design* [194] to incentivize the agents to behave truthfully when participating in a coalition formation process. Recall that mechanism design deals with settings where selfish agents possess private information and the center's goal is to implement a certain function that may depend on this private information. The center's task is complicated by the fact that agents have their own preferences over the possible outcomes, and will lie to the center if this leads to an outcome that they prefer to the truthful outcome. Thus, the agents' incentives need to be aligned with those of the center, typically by promising monetary transfers to the agents.

The use of mechanism design in coalition formation dates back to Myerson [190] (see also [191]), who presented the very first Bayesian model of coalitional games. In his model, players are assumed to know their own type, and have probabilistic beliefs about other players' types. The game outcome is determined by a mechanism implemented by some *mediator* which, given reported agent types, chooses coalitional actions and distributes the payoffs among the agents. The mediator plays an important role in preventing information leakage: in the absence of a mediator, agents would have to approach each other with proposals to form a coalition, and the mere fact of such proposal may provide additional information about the type of the proposing agent. The mediator is required to be *incentive-compatible*, i.e., the side payments to the agents should ensure truth-telling, as well as *budget-balanced*, i.e., the total transfer from the mediator to the players should be non-positive. Myerson [190] proposed an extension of the Shapley value to a large class of cooperative games with incomplete information.

In a subsequent paper [192], Myerson analyzes stability in cooperative games with incomplete information via the notion of a *blocking mediator*. After the original mediator has approached the players and assigned them payoffs, the blocking mediator can approach a subset S of the players and attempt to make a better offer: that is, he asks each $i \in S$ for his type, and based on that, he offers a distribution over coalitional actions for S and a payoff to each agent in S. If the expected payoff to each member of S is higher than that offered by the original mediator, S is said to have a *blocking plan*. It is important to note that by offering the deviation plan, the mediator may give the players additional information about the other players' types. In order to sidestep this difficulty, Myerson proposes that a blocking mediator approaches sets according to some probability distribution, and may suggest a collaboration plan even when it is not worthwhile to the players. A blocking mediator is required to be incentive compatible and budget balanced. A mediator's payoff allocation is said to be *inhibitive* if no set can find a blocking plan. Myerson proceeds to give characterizations of core mediators via conditions similar to balancedness in classic cooperative games.

Within the multiagent community, Yamamoto and Sycara [265] and Li and Sycara [169] have also proposed versions of a core concept under incomplete information, to be used with coalition formation protocols that enable group buying and group bidding activities in electronic marketplaces.

Their core versions refer to stability based on payoff sharing within a single coalition only, and not across coalitions. Li *et al.* [168] propose a different core concept referring to stability across all coalitions in an e-marketplace where the buyers come together to profit from group discounts. This core characterizes stability based on the utility of the coalitions' members as reported to some central manager. This paper also presents an *egalitarian* payoff division mechanism, in which coalition members share buying costs as evenly as possible, subject to individual rationality and budget balance constraints. Empirically, this mechanism is shown to maximize social welfare while leading to stable outcomes and incentivizing the buyers to truthfully reveal their valuations. However, these results are *only* empirical: as a matter of fact, the paper presents an impossibility result regarding the existence of an incentive compatible mechanism for such coalitional games.

7.2.1 ANONYMITY-PROOF SOLUTION CONCEPTS

Another class of coalitional games that can be usefully studied from the mechanism design perspective includes games that arise in open, anonymous environments such as the Internet. In such games, the set of agents that may choose to participate in a coalitional game is not known in advance, and, moreover, agents' identities cannot always be verified. This complicates the coalition formation process considerably, as agents acquire new means to cheat during the coalition formation process: for instance, several colluding agents may pretend to be a single agent, or, conversely, a single agent may participate in the coalition formation process under multiple identifiers; the latter type of dishonest behavior is known as *false-name manipulation* [77]. In particular, standard solution concepts such as the Shapley value and the core can be shown to be vulnerable to such manipulations. Thus, one may want to employ mechanism design techniques to encourage the agents to behave truthfully,

To address this challenge, Yokoo *et al.* [269] propose a model of coalitional games where the value of a coalition is determined in terms of *skills*, or *resources* that it possesses (rather than identifiers of its members). This idea alone does not suffice to incentivize truthful behavior, as agents may still benefit from hiding their skills. Thus, Yokoo *et al.* introduce the notion of the *anonymity-proof core*, which is robust to these manipulations. They then proceed to show that the anonymity-proof core is non-empty if a set of axioms is satisfied. Ohta *et al.* [197] later build on [269] to propose and study the *anonymity-proof Shapley value*.

7.3 OVERLAPPING AND FUZZY COALITION FORMATION

All the models of coalitional games that we have examined so far assumed that agents could belong to only one coalition at a time—i.e., that any coalition formation process results in a *partition* of agents into *disjoint coalitions*. However, in many scenarios of interest, this assumption is not applicable. Specifically, it is often natural to associate coalitions with tasks to be performed by the agents. In such situations, some agents may be involved in several tasks, and therefore may need to distribute their resources among the coalitions in which they participate. Indeed, such "overlaps" may be necessary to obtain a good outcome. As a simple e-commerce example, consider online trading agents representing individuals or virtual enterprises, and facing the challenge of allocating their owners' capital to a variety of projects (i.e., coalitions) simultaneously. There are many other examples of settings in which an agent (be it a software entity or a human) *splits his resources* (such as processing power, time, or money) among *several* tasks. These tasks, in turn, may require the participation of more than one agent: a computation may run on several servers, a software project usually involves more than one engineer, and a start-up may rely on several investors. Thus, each task corresponds to a coalition of agents, but agents' contributions to those coalitions may be fractional, and, moreover, agents can participate in several tasks at once, resulting in coalition structures with *overlapping coalitions*.

One of the earliest papers that studied coalition formation with overlapping coalitions is the work by Shehory and Kraus [236]. In the model of [236], agents have goals and capabilities— i.e., abilities to execute certain actions. To achieve their goals, the agents have to participate in coalitions, to each of which they contribute some of their capabilities. Shehory and Kraus propose heuristic algorithms that lead to the creation of overlapping coalition structures. Later, Dang *et al.* [83] examined heuristic algorithms for overlapping coalition formation to be used in surveillance multi-sensor networks. However, these papers approach coalition formation from the optimization perspective and ignore game-theoretic issues altogether.

Recently, Chalkiadakis *et al.* [71] proposed a model for overlapping coalition formation. In their model, each agent is endowed with a certain amount of resources, which he is free to distribute across multiple coalitions. The value of such (partial) coalition is determined both by the identities of agents that participate in it and the amount of resources that they contribute. Mathematically, an overlapping coalition formation game (OCF game) with n players is specified by a characteristic function $v : [0, 1]^n \to \mathbb{R}$. This function is defined on *partial coalitions*, i.e., vectors of the form $\mathbf{r} = (r_1, \ldots, r_n)$, where r_i is the fraction of agent i's resources contributed to this coalition; function v maps any such coalition $\mathbf{r}$ to a corresponding payoff. An outcome of an OCF game is an *overlapping coalition structure*, i.e., a finite collection of partial coalitions $(\mathbf{r}^j)_{j=1,\ldots,s}$ such that $\sum_{j=1}^s r_i^j \le 1$ for any player i.

Chalkiadakis *et al.* focus on the study of stability in their model. Compared to the non-overlapping setting, the stability of an overlapping coalition structure is a delicate issue: if an agent is participating in several projects at once and decides to withdraw all or some of her contributions from one of them, can she expect to continue to receive the payoff from the coalitions that were not harmed by the deviation? Chalkiadakis *et al.* propose three different stability concepts—the *conservative* core, the *refined* core, and the *optimistic* core—that correspond to three possible ways of answering this question. Briefly, under the conservative core, the deviators do not expect to get any payoffs from their coalitions with non-deviators. In contrast, in the refined core, they continue to get payoffs from coalitions not affected by the deviation. Finally, in the optimistic core, the deviators may get some payoffs from an affected coalition, as long as they continue to contribute to it, and the members of that coalition were able to regroup and focus on a different task so that each non-deviator still gets as much profit as before from that coalition. Chalkiadakis *et al.* provide a characterization of outcomes in the pessimistic core, define a notion of convexity for OCF games and prove that for convex games the pessimistic core is non-empty. They also study the complexity of core-related questions in OCF games for a simple task-based representation.

These results are extended by Zick and Elkind [270], who proposed the notion of *arbitrated core* for games with overlapping coalitions, which encompasses all three definitions of the core given in [71]. They then extend the characterization given in [71] to the arbitrated core, as well as define and study the nucleolus and two variants of Shapley value for OCF games. Theirs is not the first attempt to extend the notion of Shapley value to games with overlapping coalition: earlier, Albizuri *et al.* [5] presented an extension of Owen's value [202] (which, in turn, can be thought of as a generalization of the Shapley value) to an overlapping coalition formation setting, and provided an axiomatic characterization of their *configuration value*. However, in the model of Albizuri *et al.*, there exists no notion of resources that an agent needs to distribute across coalitions, and hence their value is different from both of the values proposed in [270].

An earlier model that also allows an agent to only contribute a part of his resources to a coalition was proposed by Aubin, who studied *fuzzy coalitional games* [9]. A player in a fuzzy game can participate in a coalition at various *levels*, and the value of a coalition S depends on the participation levels of the agents in S. A detailed exposition of fuzzy games is provided by Brânzei *et al.* [61]. Formally speaking, a fuzzy game is defined in the same way as an OCF game, i.e., by a characteristic function $v : [0, 1]^n \to \mathbb{R}$. However, the only acceptable outcome of a fuzzy game is the formation of the grand coalition along with a payoff allocation to the agents, and partial coalitions are only used to define stability: the *fuzzy core* (also referred to as the *Aubin core*) consists of all outcomes that are stable against deviations by partial ("fuzzy") coalitions. Chalkiadakis *et al.* give a detailed comparison between fuzzy games and OCF games, showing that the fuzzy core is different from all core variants considered in [71]. Fuzzy versions of other coalitional solution concepts have also been studied: Blankenburg *et al.* [54] employ the concept of the *fuzzy kernel* in their work on sequential, bilateral coalition negotiations with uncertain coalition values, and Blankenburg and Klusch [53] propose a coalition formation algorithm that is based on the *fuzzy bilateral Shapley value*.

7.4   LOGICAL APPROACHES TO COOPERATIVE GAME THEORY

Knowledge representation is a key issue in research in artificial intelligence. Knowledge representation is largely concerned with the development of logics and related formalisms to represent and reason about domains of discourse. If we take coalitional games as our domain of discourse, then a very natural question from the perspective of logic and knowledge representation is what kinds of logics might be used to represent and reason about coalitional games?

The first substantial work in this area was carried out by Pauly [205]. He developed a logic called *Coalition Logic*, which was explicitly intended for reasoning about cooperative scenarios. In Coalition Logic, the main construct is a *cooperation modality*: an expression of the form $[C]\varphi$, where C is a set of agents and φ is a logical property. The intended interpretation of $[C]\varphi$ is that C has the ability to ensure that φ is true, irrespective of what the other players (outside the coalition C) do. To give a precise meaning (i.e., a formal semantics) to Coalition Logic, Pauly used *effectivity functions*, and the notion of α-effectivity, as described earlier in this chapter (see [3]). Thus $[C]\varphi$ means that there exists a tuple of strategies, one for each member of C, such that if the players in C follow these strategies, then φ is guaranteed to result. A rich generalization of Coalition Logic was developed by Alur *et al.* [6] in their *Alternating-time Temporal Logic* (ATL), a formalism which has been widely used within the multiagent systems community for reasoning about game-like multiagent systems (see [262] for a survey).

In somewhat related work, Ågotnes *et al.* [4] developed several logics specifically intended for reasoning about NTU games, as described in chapter 5. For example, they introduced a logic with expressions such as $\langle C \rangle \lambda$, meaning that λ is a choice for coalition C (i.e., $\lambda \in v(C)$). Using this logic, they were able to characterize various solution concepts of cooperative game theory, and moreover were able to prove theorems of game theory within the logic itself.

7.5   APPLICATIONS

We have already mentioned, either explicitly or in passing, several applications of cooperative games in real-life settings, such as electronic marketplaces, virtual organizations, international policy-making, surveillance networks, and multiagent task-allocation scenarios. In this section, we will discuss a number of application domains that are of particular interest from a technological, economic, and, indeed, societal point of view.

7.5.1   COALITIONS IN COMMUNICATION NETWORKS

In recent years, communication networks have been acquiring an interesting list of epithets: *mobile*, *wireless*, and *ad hoc*. In a nutshell, communication networks increasingly rely on mobile, wireless devices acting as nodes coalescing together to allow for the routing of packets on an *ad hoc* basis. It is to be expected, therefore, that the increased autonomy of network devices leads to a need for them to act as rational decision makers. This is where game theory comes naturally into the picture, providing the means to study the strategic interactions of the network devices (agents) in this context. Indeed, ideas from cooperative game theory have recently found their way into the theory and practice of communication networks. Here we mention just a few papers representative of this trend, and refer the reader to an excellent recent review paper of Saad *et al.* [223] for more.

To begin, there is a growing literature on *strategic network formation* [145]. While most of the work on this topic focuses on the non-cooperative aspects of the network formation game (i.e., aims to establish and characterize equilibrium behavior when building the network), stability is also a major concern. For instance, Arcaute *et al.* [7] present a protocol by which the network formation process converges to a *pairwise-stable* and efficient tree network, through bilateral agreements between nodes.

Han and Poor [133] employ coalitional game theory to deal with the *curse of the boundary nodes* in packet forwarding *ad hoc* networks. Here, the problem is that nodes lying at the fringe of a network are of little value to central nodes, and thus cannot find relays for transmitting their packets, as central nodes have no incentive to perform this service for them. Han and Poor show that a central, *backbone* node of the network can reduce its power consumption by cooperating with a number of boundary nodes which act as relays for transmitting its packets. In return, boundary nodes get to send their packets through the central node, overcoming their inherent disadvantage in finding relays for their transmissions.

In mobile communication networks, the formation of (virtual) *multiple-input, multiple output (MIMO)* systems can be facilitated through the dynamic formation of coalition among single-antenna users that cooperate to share their antennas. Saad *et al.* [222] recently proposed an algorithm for the dynamic formation of such coalitions, through continuous splits and merges to account for changing transmission costs due to the mobility of the users. The algorithm has the attractive property of converging to welfare maximizing, stable outcomes, if such outcomes exist.

Finally, coalitional games have many applications in the domain of *network security*, where players may need to act together to fend off attackers. We refer the reader to the work of Grossklags *et al.* [130] for a non-cooperative game-theoretic model of network security, and recent work by Harikrishna *et al.* [134] for a cooperative game theoretic view of this problem. Other recent work on this topic includes the papers by Aziz *et al.* [23], and Bachrach and Porat [32].

7.5.2 COALITIONS IN THE ELECTRICITY GRID

Research on applications of cooperative game theory in the electricity generation and transmission domain has a long history. For instance, Contreras *et al.* [78, 79] presented a bilateral Shapley value negotiation scheme to determine how to best share the costs for the expansion of power transmission networks among collaborating rational agents. Subsequently, Contreras and Wu [80] proposed the use of the kernel solution concept for building stable coalitions to bear the costs of the transmission network expansion. Yeung *et al.* [268] employed coalitional game theory to model the trading process between market entities that generate, transmit and distribute power.

Recently, this research area has been revitalized by the vision of a "Smart Grid" [253] and the resulting creation of a robust, intelligent electricity supply network which makes efficient use of energy resources [85, 98, 155, 258, 259]. In this context, one of the main problems facing the energy supply industry is how to best achieve the utilization of the *distributed energy resources (DERs)* that, in recent years, have appeared in the electricity network. Such DERs range from electricity storage devices to small and medium capacity renewable energy generators. In principle, employing DERs to produce energy could reduce reliance on conventional power plants by as much as half [212]. Unlike conventional power plants that rely on the transmission network and are "dispatched" (i.e., called in to produce energy when needed) by the national electricity transmission network operators (termed *the Grid*), DERs lie in the distribution network and, due to their small size, they (and their capacity) are "invisible" to the Grid. Thus, they cannot be easily dispatched to meet demand. Moreover, due to their decentralized nature and small size, DERs are either invisible to the electricity market as well, or, when operating alone, usually lack the capacity, flexibility or controllability to participate in a cost-efficient way [212].

Now, the *reliability of supply* is a major concern of the Grid. It is essential that independent suppliers are reliable, since failure to meet production targets could seriously compromise the smooth operation of the network as a whole. In contrast, given the unpredictability of renewable energy sources, the DERs would usually struggle to meet power generation targets when operating alone. This normally prohibits them from striking profitable deals with the Grid, and keeps them out of the electricity market for fear of suffering penalties specified in contracts. This drives them to sign low-profit contracts with third-party market participants instead [212]. With the number of DERs expected to rise to hundreds of thousands, and, with the variable generation seen as another uncertainty to be addressed in real time through active Grid management, their integration into the electricity network is inherently problematic.

It has been recently proposed [155, 212] that this problem can be countered by creating *Virtual Power Plants (VPPs)* that aggregate DERs into the virtual equivalent of a large power station, and thus enable them to cost-effectively integrate into the market. A VPP is a broad term that represents the aggregated capabilities of a set of DERs. For example, it can be thought of as a portfolio of DERs, as an independent entity that coordinates DERs, as an external aggregator that "hires" DERs to profit from their exploitation, or—importantly—as a *coalition of DERs*. The latter perspective has been considered by Dimeas and Hatziargyriou [98], who essentially suggest an organizational structure for VPPs that makes use of interacting coalitions, and by Mihailescu *et al.* [184], who propose the use of coalition formation to build VPPs, but do not provide the details of the process or offer specific game-theoretic solutions.

The *PowerMatcher* is a decentralized system architecture that has been proposed by Kok *et al.* [155] as a means to balance demand and supply in clusters of DERs. It attempts to implement optimal supply and demand matching by organizing the DER-agents into a logical tree, assigning them roles and prescribing strategies to use in their interactions. Then, the individual agents' supply offers are aggregated in a cluster, corresponding to a VPP through the use of an *objective agent*. Such an agent has the task of implementing a "business logic" that would guide the VPP's actions. However, Kok *et al.* stop short of proposing a specific business logic.

Chalkiadakis *et al.* [73] proposed what can be seen as a detailed description of just such a logic, employing game-theoretic ideas and tools to this purpose. They describe specific mechanisms for the market-VPP interface and the interactions among VPP members. They discuss the issues surrounding the emergence of *cooperative virtual power plants (CVPPs)*, and propose a pricing mechanism for CVPPs. Their mechanism guarantees that CVPPs have the incentive to truthfully report to the Grid accurate estimates of their electricity production, and that larger rather than smaller CVPPs form; this promotes CVPP efficiency and reliability. In addition, they propose a scheme to allocate payments within the cooperative, and show that, given this scheme and the pricing mechanism, the allocation is in the core and, as such, no subset of members has a financial incentive to break away from the CVPP.

7.5.3   CORE-SELECTING AUCTIONS

Auctions are among the most important economic mechanisms used to allocate goods and procure services [161, 185]. For instance, in the USA the auctions that sell licenses to use different bands of the electromagnetic spectrum (the so-called spectrum auctions) have raised over $60 billion in revenue so far. Many practically useful auctions (including spectrum auctions) sell several items simultaneously, and a bidder's valuation for an item may depend on which other items she receives; such auctions are known as *combinatorial* [81]. Finding good mechanisms for combinatorial domains is a challenging task: while there exist auction mechanisms that elicit truthful bidding and ensure efficient allocation of the items being sold (the so-called *VCG auctions* [75, 132, 255]), these mechanisms have high communication and computational complexity, their revenue may be low and they are vulnerable to shill bidding [16]. In fact, counterintuitively, in a VCG auction there may be a group of losing bidders whose joint willingness to pay for the items exceeds the revenue collected by the seller. In other words, in the coalitional game induced by the VCG mechanism, where the set of players consists of the buyers and the seller, a coalition that consists of the seller and some of the losing bidders can profitably deviate from the outcome prescribed by the mechanism. This observation motivates the definition of *core-selecting auctions*: these are auctions whose outcome is guaranteed to be in the core of the respective coalitional game. Day and Milgrom [87] show that core-selecting auctions have a number of desirable properties: they are resistant to false-name bidding [77], and their revenue is non-decreasing in bids (which is not the case for the VCG auction). The concept of a core-selecting auction has been influential is designing practical auction protocols: for instance, a recent successful spectrum auction in the United Kingdom was designed to be core-selecting, and raised substantial revenue for the government.

7.6   RESEARCH DIRECTIONS

As a parting note, in this section we list several topics in coalitional game theory that, in our opinion, present interesting research opportunities for the immediate future.

The first one is algorithmic aspects of coalitional games in *structured* environments. These are environments that can be modeled by identifying the players with the vertices of a graph and allowing a coalition to form only if it corresponds to a connected subgraph. Intuitively, edges encode relationships between players (an edge from a to b means that a and b know each other, and/or have a communication link between them), so a disconnected subgraph corresponds to a group of players whose members cannot coordinate and hence form a functioning coalition. Many real-world domains can be modeled in this way, including, e.g., sensor or telecommunication networks.

Graph-based models have long been recognized as a natural framework for the study of coalition formation. They date back to the seminal work of Myerson [189], in which he introduced an analogue of the Shapley value for graph-restricted games, which was subsequently termed the *Myerson value*. Subsequently, Le Breton *et al.* [163] and Demange [92] analyzed stability in such games, showing that if a game is superadditive and the graph is a tree, the core is non-empty. In a follow-up paper, Demange [93] showed that this remains true even for non-superadditive games, and proposed a procedure that computes an element in the core of such a game. A fairly recent survey of graph-restricted games can be found in [254]. However, there has not been much work on computing solution concepts in graph-restricted games.

We remark that for several classes of games, including MC-nets [128, 143], skill games [30] and planning games [60], positive algorithmic results have been obtained by constructing a graph from the natural representation of the game and designing algorithms whose running time is polynomial in the number of agents as long as this graph has bounded treewidth. Thus, graph-theoretic techniques have already proved to be useful for the analysis of coalitional games.

A second—and somewhat related—direction with much research potential is studying coalitional games in conjunction with learning in *organizations*. Organizations provide natural environments for multiple layers of learning, and for transfer of knowledge among such layers. Moreover, the scalability of cooperative solutions might be enhanced due to the underlying hierarchical structure. Though there has been some initial work on this topic (see, e.g., the work of Abdallah and Lesser [2] mentioned in the chapter 6), much more can be done here, in our view.

Games with externalities, that is, partition function games (PFGs) (see section 5.2) provide a third promising research domain. We have already mentioned recent work on the coalition structure generation problem in PFGs [37, 217] as well as the question of obtaining efficient PFG representations [182]. However, the question of stability in PFG settings is still a topic of ongoing discussion in the economics literature; we refer to [154] for an overview of proposed PFG solution concepts.

We also believe that there is more room for interesting work at the intersection of coalition formation and mechanism design. Coalition formation could potentially be viewed as a decentralized mechanism design tool that allows agents to *collude* as they find appropriate when participating in mechanisms for collective decision-making or allocation of goods and services. We believe that several interesting questions exist in this context: How are the outcomes of standard mechanisms (such as, e.g., VCG) in the presence of colluding agents related to outcomes of more sophisticated mechanisms that explicitly take collusion into account? How can colluding agents share the benefits obtain from collaboration? What is the impact of the collusion on the center's utility? How do colluding agents fare *under uncertainty*? Some of these questions have been studied in the context of first-price auctions [165, 166, 167], multi-unit auctions [25], VCG auctions [28] and voting [26], but many interesting questions remain open.

Another promising topic is the study of overlapping coalition formation: While Chalkiadakis *et al.* [71] present a conceptual framework for this problem, there are many questions that were not addressed in their work, or the follow-up work of Zick and Elkind [270]. This includes identifying suitable representation formalisms for such games, devising efficient procedures for overlapping coalition formation and linking theoretical results with agent behavior encountered in real-world domains where agents form overlapping coalitions.

Finally, the design of decentralized coalition formation protocols with desirable characteristics for realistic multiagent settings is a challenging research question. For instance, what is the best course of action if communication fails in an uncertain environment? How is one to account for communication costs when forming coalitions? What is the complexity, and, in particular, the communication complexity of coalition formation protocols? These questions are far from having been settled, and, in particular, the game-theoretic aspects of interaction under uncertainty and costly communication are not well understood. The challenge, therefore, is there to be met.

Bibliography

[1] K. V. Aaditya, T. Michalak, and N. R. Jennings. Representation of coalitional games with algebraic decision diagrams (extended abstract). In *AAMAS'11: 10th International Joint Conference on Autonomous Agents and Multiagent Systems*, pages 1121–1122, 2011. Cited on page(s) 46

[2] S. Abdallah and V. Lesser. Organization-based coalition formation. In *AAMAS'04: 3rd International Joint Conference on Autonomous Agents and Multiagent Systems*, pages 1296–1297, 2004. DOI: 10.1109/AAMAS.2004.206 Cited on page(s) 105, 119

[3] J. Abdou and H. Keiding. *Effectivity Functions in Social Choice Theory*. Kluwer, 1991. Cited on page(s) 108, 115

[4] T. Ågotnes, W. van der Hoek, and M. Wooldridge. Reasoning about coalitional games. *Artificial Intelligence*, 173(1):45–79, 2009. DOI: 10.1016/j.artint.2008.08.004 Cited on page(s) 115

[5] M. J. Albizuri, J. Aurrecoechea, and J. M. Zarzuelo. Configuration values: extensions of the coalitional Owen value. *Games and Economic Behavior*, 57:1–17, 2006. DOI: 10.1016/j.geb.2005.08.016 Cited on page(s) 114

[6] R. Alur, T. A. Henzinger, and O. Kupferman. Alternating-time temporal logic. *Journal of the ACM*, 49(5):672–713, 2002. DOI: 10.1145/585265.585270 Cited on page(s) 115

[7] E. Arcaute, R. Johari, and S. Mannor. Network formation: bilateral contracting and myopic dynamics. In *WINE'07: 3rd International Workshop on Internet and Network Economics*, pages 191–207, 2007. DOI: 10.1109/TAC.2009.2024564 Cited on page(s) 115

[8] T. Arnold and U. Schwalbe. Dynamic coalition formation and the core. *Journal of Economic Behavior & Organization*, 49(3):363–380, 2002. DOI: 10.1016/S0167-2681(02)00015-X Cited on page(s) 94, 101

[9] J.-P. Aubin. Cooperative fuzzy games. *Mathematics of Operations Research*, 6(1):1–13, 1981. DOI: 10.1287/moor.6.1.1 Cited on page(s) 114

[10] J. Augustine, N. Chen, E. Elkind, A. Fanelli, N. Gravin, and D. Shiryaev. Dynamics of profit-sharing games. In *IJCAI'11: 21st International Joint Conference on Artificial Intelligence*, pages 37–42, 2011. Cited on page(s) 95

[11] R. J. Aumann. Acceptable points in general cooperative n-person games. In A. W. Tucker and L. D. Luce, editors, *Contributions to the Theory of Games, volume IV*, pages 287–324. Princeton University Press, 1959. Cited on page(s) 107, 108

[12] R. J. Aumann and J. Dréze. Cooperative games with coalition structures. *International Journal of Game Theory*, 3(4):217–237, 1974. DOI: 10.1007/BF01766876 Cited on page(s) 26

[13] R. J. Aumann and M. Maschler. The bargaining set for cooperative games. In M. Dresher, L. S. Shapley, and A. W. Tucker, editors, *Advances in Game Theory*, pages 443–447. Princeton University Press, 1964. Cited on page(s) 33

[14] R. J. Aumann and B. Peleg. Von Neumann–Morgenstern solutions to cooperative games without side payments. *Bulletin of the American Mathematical Society*, 66:173–179, 1960. DOI: 10.1090/S0002-9904-1960-10418-1 Cited on page(s) 107

[15] G. Ausiello, P. Crescenzi, G. Gambosi, V. Kann, A. Marchetti-Spaccamela, and M. Protasi. *Complexity and Approximation*. Springer-Verlag, Berlin, 1999. Cited on page(s) 62

[16] L. M. Ausubel and P. Milgrom. The lovely but lonely Vickrey auction. In P. Cramton, Y. Shoham, and R. Steinberg, editors, *Combinatorial Auctions*. MIT Press, 2006. Cited on page(s) 118

[17] R. Axelrod. *The Evolution of Cooperation*. Basic Books, 1984. Cited on page(s) 6

[18] H. Aziz, Y. Bachrach, E. Elkind, and M. S. Paterson. False-name manipulations in weighted voting games. *Journal of Artificial Intelligence Research*, 40:57–93, 2011. Cited on page(s) 57, 58

[19] H. Aziz, F. Brandt, and P. Harrenstein. Monotone cooperative games and their threshold versions. In *AAMAS'10: 9th International Joint Conference on Autonomous Agents and Multiagent Systems*, pages 1107–1114, 2010. DOI: 10.1145/1838206.1838356 Cited on page(s) 41

[20] H. Aziz, F. Brandt, and H.G. Seedig. Optimal partitions in additively separable hedonic games. In *IJCAI'11: 21st International Joint Conference on Artificial Intelligence*, pages 43–48, 2011. Cited on page(s) 77

[21] H. Aziz, F. Brandt, and H.G. Seedig. Stable partitions in additively separable hedonic games. In *AAMAS'11: 10th International Joint Conference on Autonomous Agents and Multiagent Systems*, pages 183–190, 2011. Cited on page(s) 77

[22] H. Aziz and B. de Keijzer. Complexity of coalition structure generation. In *AAMAS'11: 10th International Joint Conference on Autonomous Agents and Multiagent Systems*, pages 191–198, 2011. Cited on page(s) 87

[23] H. Aziz, O. Lachish, M. Paterson, and R. Savani. Wiretapping a hidden network. In *WINE'09: 5th International Workshop on Internet and Network Economics*, pages 438–446, 2009. DOI: 10.1007/978-3-642-10841-9-40 Cited on page(s) 109, 116

[24] H. Aziz and T. B. Sørensen. Path coalitional games. In *CoopMAS'11: 2nd Workshop on Cooperative Games in Multiagent Systems*, 2011. Cited on page(s) 109

[25] Y. Bachrach. Honor among thieves—collusion in multi-unit auctions. In *AAMAS'10: 9th International Conference on Autonomous Agents and Multiagent Systems*, pages 617–624, 2010. Cited on page(s) 119

[26] Y. Bachrach., E. Elkind, and P. Faliszewski. Coalitional voting manipulation: A game-theoretic perspective. In *IJCAI'11: 22nd International Joint Conference on Artifical Intelligence*, pages 49–54, 2011. Cited on page(s) 119

[27] Y. Bachrach, E. Elkind, R. Meir, D. Pasechnik, M. Zuckerman, J. Rothe, and J. S. Rosenschein. The cost of stability in coalitional games. In *SAGT'09: 2nd International Symposium on Algorithmic Game Theory*, pages 122–134, 2009. DOI: 10.1007/978-3-642-04645-2-12 Cited on page(s) 29, 65

[28] Y. Bachrach, P. Key, and M. Zadimoghaddam. Collusion in VCG path procurement auctions. In *WINE'10: 6th International Workshop on Internet and Network Economics*, pages 38–49, 2010. DOI: 10.1007/978-3-642-17572-5-4 Cited on page(s) 119

[29] Y. Bachrach, E. Markakis, E. Resnick, A. D. Procaccia, J. S. Rosenschein, and A. Saberi. Approximating power indices: theoretical and empirical analysis. *Autonomous Agents and Multi-Agent Systems*, 20(2):105–122, 2010. DOI: 10.1007/s10458-009-9078-9 Cited on page(s) 56

[30] Y. Bachrach, R. Meir, K. Jung, and P. Kohli. Coalitional structure generation in skill games. In *AAAI'10: 24th AAAI Conference on Artificial Intelligence*, pages 703–708, 2010. Cited on page(s) 87, 119

[31] Y. Bachrach, M. Polukarov, and N. R. Jennings. The good, the bad and the cautious: Safety level cooperative games. In *WINE'10: 6th International Workshop on Internet and Network Economics*, pages 432–443, 2010. DOI: 10.1007/978-3-642-17572-5-36 Cited on page(s) 108

[32] Y. Bachrach and E. Porat. Path disruption games. In *AAMAS'10: 9th International Conference on Autonomous Agents and Multiagent Systems*, pages 1123–1130, 2010. Cited on page(s) 116

[33] Y. Bachrach and J. S. Rosenschein. Coalitional skill games. In *AAMAS'08: 7th International Conference on Autonomous Agents and Multiagent Systems*, pages 1023–1030, 2008. Cited on page(s) 46

[34] Y. Bachrach and J. S. Rosenschein. Power in threshold network flow games. *Autonomous Agents and Multi-Agent Systems*, 18(1):106–132, 2009. DOI: 10.1007/s10458-008-9057-6 Cited on page(s) 41

[35] R. I. Bahar, E. A. Frohm, C. M. Gaona, G. D. Hachtel, E. Macii, A. Pardo, and F. Somenzi. Algebraic decision diagrams and their applications. *Formal Methods in System Design*, 10:171–206, 1997. DOI: 10.1023/A:1008699807402 Cited on page(s) 46

[36] C. Ballester. NP-competeness in hedonic games. *Games and Economic Behavior*, 49:1–30, 2004. DOI: 10.1016/j.geb.2003.10.003 Cited on page(s) 76, 77

[37] B. Banerjee and L. Kraemer. Coalition structure generation in multi-agent systems with mixed externalities. In *AAMAS'10: 9th International Conference on Autonomous Agents and Multiagent Systems*, pages 175–182, 2010. Cited on page(s) 92, 119

[38] B. Banerjee and S. Sen. Selecting partners. In *Agents'00: 4th International Conference on Autonomous Agents*, pages 261–262, 2000. DOI: 10.1145/336595.337478 Cited on page(s) 97

[39] S. Banerjee, H. Konishi, and T. Sönmez. Core in a simple coalition formation game. *Social Choice and Welfare*, 18:135–153, 2001. DOI: 10.1007/s003550000067 Cited on page(s) 75

[40] J. F. Banzhaf. Weighted voting doesn't work: A mathematical analysis. *Rutgers Law Review*, 19:317–343, 1965. Cited on page(s) 22

[41] E. T. Bell. Exponential numbers. *American Mathematical Monthly*, 41:411–419, 1934. DOI: 10.2307/2300300 Cited on page(s) 88

[42] R. Bellman. A problem in the sequential design of experiments. *Sankhya*, 16:221–229, 1956. Cited on page(s) 102

[43] R. E. Bellman. *Dynamic Programming*. Princeton University Press, 1957. Cited on page(s) 104

[44] B. Bernheim, B. Peleg, and M. Whinston. Coalition-proof Nash equilibria I: Concepts. *Journal of Economic Theory*, 42(1):1–12, 1987. DOI: 10.1016/0022-0531(87)90100-1 Cited on page(s) 108

[45] J. Bilbao, J. R. Fernández, A. Losada, and J. López. Generating functions for computing power indices efficiently. *TOP: An Official Journal of the Spanish Society of Statistics and Operations Research*, 8(2):191–213, 2000. DOI: 10.1007/BF02628555 Cited on page(s) 57

[46] J. M. Bilbao, J. R. Fernández, N. Jiminéz, and J. J. López. Voting power in the European Union enlargement. *European Journal of Operational Research*, 143:181–196, 2002. DOI: 10.1016/S0377-2217(01)00334-4 Cited on page(s) 66, 69

[47] K. Binmore. *Fun and Games: A Text on Game Theory*. D. C. Heath and Company: Lexington, MA, 1992. Cited on page(s) 5

[48] K. Binmore. *Game Theory and the Social Contract Volume 1: Playing Fair*. MIT Press, 1994. Cited on page(s) 110

[49] K. Binmore. *Game Theory and the Social Contract Volume 2: Just Playing*. MIT Press, 1998. Cited on page(s) 110

[50] K. Binmore. *Game Theory: A Very Short Introduction*. Oxford University Press, 2007. Cited on page(s) 7, 10

[51] C. G. Bird. On cost allocation for a spanning tree: A game theory approach. *Networks*, 6:335–350, 1976. DOI: 10.1002/net.3230060404 Cited on page(s) 42

[52] B. Blankenburg, R. K. Dash, S. D. Ramchurn, M. Klusch, and N. R. Jennings. Trusted kernel-based coalition formation. In *AAMAS'05: 4th International Joint Conference on Autonomous Agents and Multiagent Systems*, pages 989–996, 2005. DOI: 10.1145/1082473.1082623 Cited on page(s) 97

[53] B. Blankenburg and M. Klusch. BSCA-F: Efficient fuzzy valued stable coalition forming among agents. In *IAT'05: 2005 IEEE/WIC/ACM International Conference on Intelligent Agent Technology*, pages 732–738, 2005. DOI: 10.1109/IAT.2005.48 Cited on page(s) 114

[54] B. Blankenburg, M. Klusch, and O. Shehory. Fuzzy kernel-stable coalitions between rational agents. In *AAMAS'03: 2nd International Conference on Autonomous Agents and Multiagent Systems*, pages 9–16, 2003. DOI: 10.1145/860575.860578 Cited on page(s) 114

[55] A. Bogomolnaia and M. O. Jackson. The stability of hedonic coalition structures. *Games and Economic Behavior*, 38:201–230, 2002. DOI: 10.1006/game.2001.0877 Cited on page(s) 75, 76

[56] S. Bolus. Power indices of simple games and vector-weighted majority games by means of binary decision diagrams. *European Journal of Operational Research*, 210(2):258–272, 2011. DOI: 10.1016/j.ejor.2010.09.020 Cited on page(s) 46

[57] O. Bondareva. Some applications of linear programming methods to the theory of cooperative games (in Russian). *Problemy Kybernetiki*, 10:119–139, 1963. Cited on page(s) 31

[58] E. Bonzon, M.-C. Lagasquie-Schiex, J. Lang, and B. Zanuttini. Boolean games revisited. In *ECAI'06: 17th European Conference on Artificial Intelligence*, pages 265–269, 2006. Cited on page(s) 83

[59] P. Borm, H. Hamers, and R. Hendrickx. Operations research games: A survey. *TOP: An Official Journal of the Spanish Society of Statistics and Operations Research*, 9:139–199, 2001. DOI: 10.1007/BF02579075 Cited on page(s) 38

[60] R. I. Brafman, C. Domshlak, Y. Engel, and M. Tennenholtz. Transferable utility planning games. In *AAAI'10: 24th AAAI Conference on Artificial Intelligence*, pages 709–714, 2010. Cited on page(s) 119

[61] R. Brânzei, D. Dimitrov, and S. Tijs. *Models in cooperative game theory*. Springer, 2005. Cited on page(s) 10, 114

[62] Simina Brânzei and Kate Larson. Coalitional affinity games and the stability gap. In *IJCAI'09: 20th International Joint Conference on Artificial Intelligence*, pages 79–84, 2011. Cited on page(s) 78

[63] Simina Brânzei and Kate Larson. Social distance games. In *IJCAI'11: 21st International Joint Conference on Artificial Intelligence*, pages 91–96, 2011. Cited on page(s) 78

[64] K. Cechlarova and J. Haidukova. Computational complexity of stable partitions with $\mathcal{B}$-preferences. *International Journal of Game Theory*, 31:353–364, 2002. Cited on page(s) 78

[65] K. Cechlarova and A. Romero-Medina. Stability in coalition formation games. *International Journal of Game Theory*, 29:487–494, 2001. DOI: 10.1007/s001820000053 Cited on page(s) 78

[66] G. Chalkiadakis. *A Bayesian Approach to Multiagent Reinforcement Learning and Coalition Formation under Uncertainty*. PhD thesis, Graduate Dept. of Computer Science, University of Toronto, 2007. Cited on page(s) 98, 100, 101, 103, 109

[67] G. Chalkiadakis and C. Boutilier. Bayesian reinforcement learning for coalition formation under uncertainty. In *AAMAS'04: 3rd International Conference on Autonomous Agents and Multiagent Systems*, pages 1090–1097, 2004. DOI: 10.1109/AAMAS.2004.74 Cited on page(s) 98, 101, 103

[68] G. Chalkiadakis and C. Boutilier. Coalitional bargaining with agent type uncertainty. In *IJCAI'07: 20th International Joint Conference on Artificial Intelligence*, pages 1227–1232, 2007. Cited on page(s) 98, 101, 103

[69] G. Chalkiadakis and C. Boutilier. Sequentially optimal repeated coalition formation under uncertainty. *Journal of Autonomous Agents and Multi-Agent Systems*, 2011. Cited on page(s) 101, 103, 104

[70] G. Chalkiadakis, E. Elkind, and N. R. Jennings. Simple coalitional games with beliefs. In *IJCAI'09: 21st International Joint Conference on Artifical Intelligence*, pages 85–90, 2009. Cited on page(s) 97

[71] G. Chalkiadakis, E. Elkind, E. Markakis, M. Polukarov, and N. R. Jennings. Cooperative games with overlapping coalitions. *Journal of Artificial Intelligence Research (JAIR)*, 39:179–216, 2010. Cited on page(s) 113, 114, 119

[72] G. Chalkiadakis, E. Markakis, and C. Boutilier. Coalition formation under uncertainty: Bargaining equilibria and the Bayesian core stability concept. In *AAMAS'07: 6th International Conference on Autonomous Agents and Multiagent Systems*, pages 400–407, 2007. DOI: 10.1145/1329125.1329203 Cited on page(s) 98, 109

[73] G. Chalkiadakis, V. Robu, R. Kota, A. Rogers, and N. R. Jennings. Cooperatives of distributed energy resources for efficient virtual power plants. In *AAMAS'11: 10th International Conference on Autonomous Agents and Multiagent Systems*, pages 787–794, 2011. Cited on page(s) 117

[74] K. Chatterjee, B. Dutta, and K. Sengupta. A noncooperative theory of coalitional bargaining. *Review of Economic Studies*, 60:463–477, 1993. DOI: 10.2307/2298067 Cited on page(s) 93

[75] E. H. Clarke. Multipart pricing of public goods. *Public Choice*, 11(1):17–33, 1971. DOI: 10.1007/BF01726210 Cited on page(s) 118

[76] V. Conitzer and T. Sandhom. Complexity of constructing solutions in the core based on synergies among coalitions. *Artificial Intelligence*, 170(6–7):607–619, 2006. DOI: 10.1016/j.artint.2006.01.005 Cited on page(s) 45

[77] V. Conitzer and M. Yokoo. Using mechanism design to prevent false-name manipulations. *AI Magazine*, 31(4):65–78, 2010. Cited on page(s) 112, 118

[78] J. Contreras, M. Klusch, and J. Yen. Coalition formation in a power transmission planning environment. In *PAAM'97: 2nd International Conference on Practical Applications of Multiagent Systems*, 1997. Cited on page(s) 116

[79] J. Contreras, M. Klusch, and J. Yen. Multi-agent coalition formation in power transmission planning: A bilateral Shapley value approach. In *AIPS'98: 4th International Conference on Artificial Intelligence in Planning Systems*, pages 19–26, 1998. DOI: 10.1109/HICSS.1998.655300 Cited on page(s) 116

[80] J. Contreras and F. F. Wu. A kernel-oriented algorithm for transmission expansion planning. *IEEE Transactions on Power Systems*, 15(4):1434–1440, 2000. DOI: 10.1109/59.898124 Cited on page(s) 116

[81] P. Cramton, Y. Shoham, and R. Steinberg, editors. *Combinatorial Auctions*. MIT press, 2006. Cited on page(s) 118

[82] I. Curiel. *Cooperative game theory and applications*. Kluwer, 1997. Cited on page(s) 10

[83] V. D. Dang, R. K. Dash, A. Rogers, and N. R. Jennings. Overlapping coalition formation for efficient data fusion in multi-sensor networks. In *AAAI'06: 21st National Conference on Artificial Intelligence*, pages 635–640, 2006. Cited on page(s) 113

[84] V. D. Dang and N. R. Jennings. Generating coalition structures with finite bound from the optimal guarantees. In *AAMAS'04: 3rd International Conference on Autonomous Agents and Multiagent Systems*, pages 564–571, 2004. DOI: 10.1109/AAMAS.2004.132 Cited on page(s) 91

[85] E. M. Davidson, M. J. Dolan, S. D. J. McArthur, and G. W. Ault. The use of constraint programming for the autonomous management of power flows. In *15th International Conference on Intelligent System Applications to Power Systems*, pages 1–7, 2009. DOI: 10.1109/ISAP.2009.5352952 Cited on page(s) 116

[86] M. Davis and M. Maschler. The kernel of a cooperative game. *Naval Research Logistics Quarterly*, 12:223–259, 1965. DOI: 10.1002/nav.3800120303 Cited on page(s) 33, 34

[87] R. Day and P. Milgrom. Core-selecting package auctions. *International Journal of Game Theory*, 36(1–2):393–407, 2008. DOI: 10.1007/s00182-007-0100-7 Cited on page(s) 118

[88] N. G. de Bruijn. *Asymptotic Methods in Analysis.* Dover, 1981. Cited on page(s) 88

[89] M. H. De Groot. *Optimal Statistical Decisions.* McGraw-Hill, 1970. Cited on page(s) 102

[90] R. Dearden, N. Friedman, and D. Andre. Model based Bayesian exploration. In *UAI'99: 15th Conference on Uncertainty in Artificial Intelligence*, pages 150–159, 1999. Cited on page(s) 102

[91] V. G. Deineko and G. J. Woeginger. On the dimension of simple monotonic games. *European Journal of Operational Research*, 170(1):315–318, 2006. DOI: 10.1016/j.ejor.2004.09.038 Cited on page(s) 68, 69

[92] G. Demange. Intermediate preferences and stable coalition structures. *Journal of Mathematical Economics*, 23:45–58, 1994. DOI: 10.1016/0304-4068(94)90035-3 Cited on page(s) 118

[93] G. Demange. On group stability in hierarchies and networks. *Journal of Political Economy*, 112(4):754–778, 2004. DOI: 10.1086/421171 Cited on page(s) 118

[94] X. Deng. Combinatorial optimization games. In C. A. Floudas and P. M. Pardalos, editors, *Encyclopedia of Optimization*, pages 387–391. Springer, 2009. Cited on page(s) 43

[95] X. Deng, Q. Fang, and X. Sun. Finding nucleolus of flow game. In *SODA'06: 17th ACM-SIAM Symposium on Discrete Algorithms*, pages 124–131, 2006. DOI: 10.1145/1109557.1109572 Cited on page(s) 33, 41

[96] X. Deng, T. Ibaraki, and H. Nagamochi. Algorithmic aspects of the core of combinatorial optimization games. *Mathematics of Operations Research*, 24(3):751–766, 1999. DOI: 10.1287/moor.24.3.751 Cited on page(s) 42, 43, 108

[97] X. Deng and C. Papadimitriou. On the complexity of cooperative solution concepts. *Mathematics of Operations Research*, 19(2):257–266, 1994. DOI: 10.1287/moor.19.2.257 Cited on page(s) 38, 39, 55

[98] A. Dimeas and N. Hatziargyriou. Agent based control of virtual power plants. In *14th International Conference on Intelligent System Applications to Power Systems*, pages 1–6, 2007. DOI: 10.1109/ISAP.2007.4441671 Cited on page(s) 116, 117

[99] D. Dimitrov, P. Borm, R. Hendrickx, and S. Sung. Simple priorities and core stability in hedonic games. *Social Choice and Welfare*, 26(2):421–433, 2006. DOI: 10.1007/s00355-006-0104-4 Cited on page(s) 78

[100] T. S. H. Driessen. *Cooperative Games, Solutions and Applications*. Kluwer, 1988. Cited on page(s) 10

[101] M. O. Duff. *Optimal Learning: Computational procedures for Bayes-adaptive Markov decision processes*. PhD thesis, Department of Computer Science, University of Massachusetts, Amherst, 2002. Cited on page(s) 102

[102] P. E. Dunne, S. Kraus, E. Manisterski, and M. Wooldridge. Solving coalitional resource games. *Artificial Intelligence*, pages 20–50, 2010. DOI: 10.1016/j.artint.2009.09.005 Cited on page(s) 83

[103] P. E. Dunne, S. Kraus, W. van der Hoek, and M. Wooldridge. Cooperative Boolean games. In *AAMAS'08: 7th International Joint Conference on Autonomous Agents and Multiagent Systems*, pages 1015–1022, 2008. Cited on page(s) 83

[104] E. Einy, R. Holzman, and D. Monderer. On the least core and the Mas-Colell bargaining set. *Games and Economic Behavior*, 28:181–188, 1999. DOI: 10.1006/game.1998.0694 Cited on page(s) 34

[105] E. Elkind, G. Chalkiadakis, and N. R. Jennings. Coalition structures in weighted voting games. In *ECAI'08: 18th European Conference on Artificial Intelligence*, pages 393–397, 2008. Cited on page(s) 58, 59

[106] E. Elkind, L. Goldberg, P. Goldberg, and M. Wooldridge. On the computational complexity of weighted voting games. *Annals of Mathematics and Artificial Intelligence*, 56(2):109–131, 2009. DOI: 10.1007/s10472-009-9162-5 Cited on page(s) 60, 65

[107] E. Elkind, L. A. Goldberg, P. Goldberg, and M. Wooldridge. A tractable and expressive class of marginal contribution nets and its applications. *Mathematical Logic Quarterly*, 55(4):362–376, 2009. DOI: 10.1002/malq.200810021 Cited on page(s) 44, 45

[108] E. Elkind, L. A. Goldberg, P. W. Goldberg, and M. Wooldridge. On the dimensionality of voting games. In *AAAI'08: 23rd AAAI Conference of Artificial Intelligence*, pages 69–74, 2008. Cited on page(s) 68, 69, 70

[109] E. Elkind and D. Pasechnik. Computing the nucleolus of weighted voting games. In *SODA'09: 20th ACM-SIAM Symposium on Discrete Algorithms*, pages 327–335, 2009. Cited on page(s) 33, 65

[110] E. Elkind and M. Wooldridge. Hedonic coalition nets. In *AAMAS'09: 8th International Conference on Autonomous Agents and Multiagent Systems*, 2009. Cited on page(s) 78, 79

[111] R. Evans. Coalitional bargaining with competition to make offers. *Games and Economic Behavior*, 19:211–220, 1997. DOI: 10.1006/game.1997.0553 Cited on page(s) 93, 94

[112] U. Faigle, W. Kern, S. P. Fekete, and W. Hochstättler. On the complexity of testing membership in the core of min-cost spanning tree games. *International Journal of Game Theory*, 26:361–366, 1997. DOI: 10.1007/BF01263277 Cited on page(s) 42

[113] U. Faigle, W. Kern, and J. Kuipers. Computing the nucleolus of min-cost spanning tree games is NP-hard. *International Journal of Game Theory*, 27:443–450, 1998. DOI: 10.1007/s001820050083 Cited on page(s) 42

[114] P. Faliszewski, E. Elkind, and M. Wooldridge. Boolean combinations of weighted voting games. In *AAMAS'09: 8th International Joint Conference on Autonomous Agents and Multiagent Systems*, pages 185–192, 2009. DOI: 10.1145/1558013.1558039 Cited on page(s) 70

[115] P. Faliszewski and L. Hemaspaandra. The complexity of power-index comparison. *Theoretical Computer Science*, 410(1):101–107, 2009. DOI: 10.1016/j.tcs.2008.09.034 Cited on page(s) 55

[116] S. S. Fatima, M. Wooldridge, and N. R. Jennings. A linear approximation method for the Shapley value. *Artificial Intelligence Journal*, 172(14):1673–1699, 2008. DOI: 10.1016/j.artint.2008.05.003 Cited on page(s) 57

[117] D. Felsenthal and M. Machover. *The Measurement of Voting Power*. Edward Elgar Publishing, Cheltenham, UK, 1998. Cited on page(s) 22, 54, 57

[118] D. Felsenthal and M. Machover. A priori voting power: What is it all about? *Political Studies Review*, 2:1–23, 2004. DOI: 10.1111/j.1478-9299.2004.00001.x Cited on page(s) 23

[119] D. Fudenberg and D. Levine. *The Theory of Learning in Games*. MIT Press, 1998. Cited on page(s) 102

[120] M. Gairing and R. Savani. Computing stable outcomes in hedonic games. In *SAGT'10: 3rd International Symposium on Algorithmic Game Theory*, pages 174–185, 2010. DOI: 10.1007/978-3-642-16170-4-16 Cited on page(s) 77

[121] M. Gairing and R. Savani. Computing stable outcomes in hedonic games with voting-based deviations. In *AAMAS'11: 10th International Conference on Autonomous Agents and Multiagent Systems*, pages 559–566, 2011. Cited on page(s) 77

[122] M. R. Garey and D. S. Johnson. *Computers and Intractibility*. W. H. Freeman and Company, 1979. Cited on page(s) xvi, 51, 60, 69

[123] A. Gerber. *Flexible Cooperation Between Autonomous Agents in Dynamic Environments*. PhD thesis, Saarland University, Germany, 2005. Cited on page(s) 105

[124] D. B. Gillies. Solutions to general non-zero-sum games. In A. W. Tucker and L. D. Luce, editors, *Contributions to the Theory of Games, volume IV*, pages 47–85. Princeton University Press, 1959. Cited on page(s) 23

[125] M. X. Goemans and M. Skutella. Cooperative facility location games. *Journal of Algorithms*, 50:194–214, 2004. DOI: 10.1016/S0196-6774(03)00098-1 Cited on page(s) 43

[126] D. Granot and F. Granot. On some network flow games. *Mathematics of Operations Research*, 17(4):792–841, 1992. DOI: 10.1287/moor.17.4.792 Cited on page(s) 41, 42

[127] D. Granot and G. Huberman. Minimum cost spanning tree games. *Mathematical Programming*, 21:1–18, 1981. DOI: 10.1007/BF01584227 Cited on page(s) 42

[128] G. Greco, E. Malizia, L. Palopoli, and F. Scarcello. On the complexity of core, kernel, and bargaining set. *Artificial Intelligence*, 175:1877–1910, 2011. DOI: 10.1016/j.artint.2011.06.002 Cited on page(s) 39, 45, 47, 119

[129] G. Greco, E. Malizia, L. Palopoli, and F. Scarcello. On the complexity of the core over coalition structures. In *IJCAI'11: 22nd International Joint Conference on Artifical Intelligence*, pages 216–221, 2011. Cited on page(s) 25, 47, 59

[130] J. Grossklags, N. Christin, and J. Chuang. Secure or insure? a game-theoretic analysis of information security games. In *WWW'08: 17th International World Wide Web Conference*, pages 209–218, 2008. DOI: 10.1145/1367497.1367526 Cited on page(s) 116

[131] M. Grötschel, L. Lovász, and A. Schrijver. *Geometric algorithms and combinatorial optimization*, volume 2 of *Algorithms and Combinatorics*. Springer-Verlag, Berlin, second edition, 1993. Cited on page(s) 61

[132] T. Groves. Incentives in teams. *Econometrica*, 41(4):617–631, 1973. DOI: 10.2307/1914085 Cited on page(s) 118

[133] Z. Han and V. Poor. Coalition games with cooperative transmission: A cure for the curse of boundary nodes in selfish packet-forwarding wireless networks. *IEEE Transactions on Communications*, 57(1):203–213, 2009. DOI: 10.1109/TCOMM.2009.0901.060661 Cited on page(s) 115

[134] N. Harikrishna, V. Venkatanathan, and C. Pandu Rangan. Towards a cooperative defense model against network security attacks. In *WEIS'10: 9th Workshop on the Economics of Information Security*, 2010. Cited on page(s) 116

[135] P. Harrenstein. *Logic in Conflict*. PhD thesis, Utrecht University, 2004. Cited on page(s) 83

[136] P. Harrenstein, W. van der Hoek, J.-J. Meyer, and C. Witteveen. Boolean games. In *TARK'01: 8th Conference on Theoretical Aspects of Rationality and Knowledge*, pages 287–298, 2001. Cited on page(s) 83

[137] S. Hart. A comparison of non-transferable utility values. *Theory and Decision*, 56:35–46, 2004. DOI: 10.1007/s11238-004-5633-7 Cited on page(s) 74

[138] S. Hart and A. Mas-Colell. Potencial, value and consistency. *Econometrica*, 57(3):589–614, 1989. DOI: 10.2307/1911054 Cited on page(s) 109

[139] S. Hart and A. Mas-Colell. Bargaining and value. *Econometrica*, 64(2):357–380, 1996. DOI: 10.2307/2171787 Cited on page(s) 93, 94

[140] M. Hoefer. Strategic cooperation in cost sharing games. In *WINE'10: 6th Workshop on Internet and Network Economics*, pages 258–269, 2010. DOI: 10.1007/978-3-642-17572-5-21 Cited on page(s) 108

[141] R. Howard. Information value theory. *IEEE Transactions on Systems Science and Cybernetics*, 2(1):22–26, 1966. DOI: 10.1109/TSSC.1966.300074 Cited on page(s) 104

[142] R. Ichimura, Y. Sakurai, S. Ueda, A. Iwasaki, M. Yokoo, and S. Minato. Compact representation scheme of coalitional games based on multi-terminal zero-suppressed binary decision diagrams. In *WSCAI'11: IJCAI'11 Workshop on Social Choice and Artificial Intelligence*, pages 40–45, 2011. Cited on page(s) 46

[143] S. Ieong and Y. Shoham. Marginal contribution nets: A compact representation scheme for coalitional games. In *ACM-EC'05: 6th ACM Conference on Electronic Commerce*, pages 193–202, 2005. DOI: 10.1145/1064009.1064030 Cited on page(s) 43, 44, 45, 78, 119

[144] S. Ieong and Y. Shoham. Bayesian coalitional games. In *AAAI'08: 23rd AAAI Conference on Artificial Intelligence*, pages 95–100, 2008. Cited on page(s) 98, 101

[145] M. Jackson. A survey of models of network formation: Stability and efficiency. In G. Demange and M. Wooders, editors, *Group Formation in Economics: Networks, Clubs and Coalitions*, 2004. Cited on page(s) 115

[146] K. Jain and M. Mahdian. Cost sharing. In N. Nisan, T. Roughgarden, E. Tardos, and V. V. Vazirani, editors, *Algorithmic game theory*, pages 385–410. Cambridge University Press, 2007. Cited on page(s) 42

[147] L. P. Kaelbling, M. L. Littman, and A. R. Cassandra. Planning and acting in partially observable stochastic domains. *Artificial Intelligence*, 101:99–134, 1998. DOI: 10.1016/S0004-3702(98)00023-X Cited on page(s) 104

[148] E. Kalai and E. Zemel. Generalized network problems yielding totally balanced games. *Operations Research*, 30(5):998–1008, 1982. DOI: 10.1287/opre.30.5.998 Cited on page(s) 40

[149] E. Kalai and E. Zemel. Totally balanced games and games of flow. *Mathematics of Operations Research*, 7(3):476–478, 1982. DOI: 10.1287/moor.7.3.476 Cited on page(s) 40

[150] W. Kern and D. Paulusma. Matching games: the least core and the nucleolus. *Mathematics of Operations Research*, 28(2):294–308, 2003. DOI: 10.1287/moor.28.2.294.14477 Cited on page(s) 33, 42

[151] S. Ketchpel. Forming coalitions in the face of uncertain rewards. In *AAAI'94: 12th National Conference on Artificial Intelligence*, pages 414–419, 1994. Cited on page(s) 96

[152] H. Kitano and S. Tadokoro. RoboCup rescue: A grand challenge for multiagent and intelligent systems. *AI Magazine*, 22(1):39–52, 2001. DOI: 10.1109/ICMAS.2000.858425 Cited on page(s) 92

[153] M. Klusch and A. Gerber. Dynamic coalition formation among rational agents. *IEEE Intelligent Systems*, 17(3):42–47, 2002. DOI: 10.1109/MIS.2002.1005630 Cited on page(s) 105

[154] L. Koczy. A recursive core for partition function form games. *Theory and Decision*, 63(1):41–51, 2007. DOI: 10.1007/s11238-007-9030-x Cited on page(s) 119

[155] K. Kok, M. Scheepers, and R. Kamphuis. Intelligence in electricity networks for embedding renewables and distributed generation. In R. R. Negenborn, Z. Lukszo, and J. Hellendoorn, editors, *Intelligent Infrastructures*. Springer, 2009. Cited on page(s) 116, 117

[156] H. Konishi and D. Ray. Coalition formation as a dynamic process. *Journal of Economic Theory*, 110:1–41, 2003. DOI: 10.1016/S0022-0531(03)00004-8 Cited on page(s) 95

[157] A. Kopelowitz. Computation of the kernels of simple games and the nucleolus of *n*-person games. Technical report, Research Program in Game Theory and Mathematical Economics, Hebrew University, 1967. Cited on page(s) 33

[158] S. Kraus, O. Shehory, and G. Taase. Coalition formation with uncertain heterogeneous information. In *AAMAS'03: 2nd International Conference on Autonomous Agents and Multiagent Systems*, pages 1–8, 2003. DOI: 10.1145/860575.860577 Cited on page(s) 96, 97

[159] S. Kraus, O. Shehory, and G. Taase. The advantages of compromising in coalition formation with incomplete information. In *AAMAS'04: 3rd International Conference on Autonomous Agents and Multiagent Systems*, pages 588–595, 2004. DOI: 10.1109/AAMAS.2004.259 Cited on page(s) 96, 97

[160] S. Kraus, J. Wilkenfeld, and G. Zlotkin. Multiagent negotiation under time constraints. *Artificial Intelligence*, 75(2):297–345, 1995. DOI: 10.1016/0004-3702(94)00021-R Cited on page(s) 96

[161] V. Krishna. *Auction Theory*. Academic Press, 2002. Cited on page(s) 118

[162] A. Laruelle and F. Valenciano. A critical reappraisal of some voting power paradoxes. *Public Choice*, 125:17–41, 2005. DOI: 10.1007/s11127-005-3408-5 Cited on page(s) 57

[163] M. Le Breton, G. Owen, and S. Weber. Strongly balanced cooperative games. *International Journal of Game Theory*, 20:419–427, 1992. DOI: 10.1007/BF01271134 Cited on page(s) 118

[164] D. Leech. Computing power indices for large voting games. *Journal of Management Science*, 49(6):831–837, 2003. DOI: 10.1287/mnsc.49.6.831.16024 Cited on page(s) 57

[165] K. Leyton-Brown, Y. Shoham, and M. Tennenholtz. Bidding clubs: institutionalized collusion in auctions. In *EC'00: 2nd ACM Conference on Electronic Commerce*, pages 253–259, 2000. DOI: 10.1145/352871.352899 Cited on page(s) 119

[166] K. Leyton-Brown, Y. Shoham, and M. Tennenholtz. Bidding clubs in first-price auctions. In *AAAI'02: 18th National Conference on Artificial Intelligence*, pages 373–378, 2002. Cited on page(s) 119

[167] K. Leyton-Brown, M. Tennenholtz, N. A. R. Bhat, and Y. Shoham. A bidding ring protocol for first-price auctions. Technical Report TR-2008-10, University of British Columbia, August 2008. Cited on page(s) 119

[168] C. Li, S. Chawla, U. Rajan, and K. Sycara. Mechanisms for coalition formation and cost sharing in an electronic marketplace. In *ICEC'03: 5th International Conference on Electronic Commerce*, pages 68–77, 2003. DOI: 10.1145/948005.948015 Cited on page(s) 112

[169] C. Li and K. Sycara. Algorithms for combinatorial coalition formation and payoff division in an e-marketplace. In *AAMAS'02: 1st International Conference on Autonomous Agents and Multiagent Systems*, pages 120–127, 2002. DOI: 10.1145/544741.544771 Cited on page(s) 111

[170] W. F. Lucas. A game with no solution. *Bulletin of the American Mathematical Society*, 74:237–239, 1968. DOI: 10.1090/S0002-9904-1968-11901-9 Cited on page(s) 35

[171] I. Mann and L. S. Shapley. Values of large games, IV: Evaluating the electoral college by Montecarlo techniques. *The Rand Corporation*, 1960. Cited on page(s) 56

[172] I. Mann and L. S. Shapley. Values of large games, VI: Evaluating the electoral college exactly. *The Rand Corporation*, 1962. Cited on page(s) 57

[173] A. Mas-Colell. An equivalence theorem for a bargaining set. *Journal of Mathematical Economics*, 18:129–139, 1989. DOI: 10.1016/0304-4068(89)90017-7 Cited on page(s) 33

[174] A. Mas-Colell, M. Whinston, and J.R. Green. *Microeconomic Theory*. Oxford University Press, 1995. Cited on page(s) 5, 10, 93, 101

[175] M. Maschler and G. Owen. The consistent Shapley value for games without side payments. In R. Selten, editor, *Rational Interaction*, pages 5–12. Springer Verlag, 1992. Cited on page(s) 74

[176] M. Maschler, B. Peleg, and L. S. Shapley. Geometric properties of the kernel, nucleolus, and related solution concepts. *Mathematics of Operations Research*, 4:303–338, 1979. DOI: 10.1287/moor.4.4.303 Cited on page(s) 28, 32

[177] T. Matsui and Y. Matsui. A survey of algorithms for calculating power indices of weighted majority games. *Journal of the Operations Research Society of Japan*, 43(1):71–86, 2000. DOI: 10.1016/S0453-4514(00)88752-9 Cited on page(s) 55

[178] Y. Matsui and T. Matsui. NP-completeness for calculating power indices of weighted majority games. *Theoretical Computer Science*, 263(1–2):305–310, 2001. DOI: 10.1016/S0304-3975(00)00251-6 Cited on page(s) 55

[179] N. Megiddo. Computational complexity of the game theory approach to cost allocation on a tree. *Mathematics of Operations Research*, 3(3):189–196, 1978. DOI: 10.1287/moor.3.3.189 Cited on page(s) 12

[180] S. Merrill. Approximations to the Banzhaf index. *American Mathematical Monthly*, 89:108–110, 1982. DOI: 10.2307/2320926 Cited on page(s) 57

[181] T. Michalak, T. Rahwan, D. Marciniak, M. Szamotulski, and N. R. Jennings. Computational aspects of extending the Shapley value to coalitional games with externalities. In *ECAI'10: 19th European Conference on Artificial Intelligence*, pages 197–202, 2010. Cited on page(s) 85

[182] T. Michalak, T. Rahwan, J. Sroka, A. Dowell, M. Wooldridge, P. McBurney, and N. R. Jennings. On representing coalitional games with externalities. In *EC'09: 10th ACM Conference on Electronic Commerce*, pages 11–20, 2009. DOI: 10.1145/1566374.1566377 Cited on page(s) 85, 119

[183] T. Michalak, J. Sroka, T. Rahwan, M. Wooldridge, P. McBurney, and N. R. Jennings. A distributed algorithm for anytime coalition structure generation. In *AAMAS'10: 9th International Conference on Autonomous Agents and Multiagent Systems*, pages 1007–1014, 2010. Cited on page(s) 85, 92

[184] R. C. Mihailescu, M. Vasirani, and S. Ossowski. Towards agent-based virtual power stations via multi-level coalition formation. In *ATES'10: 1st International Workshop on Agent Technologies for Energy Systems*, pages 107–108, 2010. Cited on page(s) 117

[185] P. Milgrom. *Putting Auction Theory to Work*. Cambridge University Press, 2004. Cited on page(s) 118

[186] B. Moldovanu and E. Winter. Order independent equilibria. *Games and Economic Behavior*, 9:21–34, 1995. DOI: 10.1006/game.1995.1003 Cited on page(s) 93, 94

[187] H. Moulin and B. Peleg. Cores of effectivity functions and implementation theory. *Journal of Mathematical Economics*, 10:115–145, 1982. DOI: 10.1016/0304-4068(82)90009-X Cited on page(s) 108

[188] S. Muroga. *Threshold Logic and its Applications*. John Wiley and Sons, 1971. Cited on page(s) 50

[189] R. Myerson. Graphs and cooperation in games. *Mathematics of Operations Research*, 2(3):225–229, 1977. DOI: 10.1287/moor.2.3.225 Cited on page(s) 118

[190] R. B. Myerson. Cooperative games with incomplete information. *International Journal of Game Theory*, 13:69–86, 1984. DOI: 10.1007/BF01769817 Cited on page(s) 98, 111

[191] R. B. Myerson. *Game Theory: Analysis of Conflict*. Harvard University Press, 1991. Cited on page(s) 5, 10, 93, 111

[192] R. B. Myerson. Virtual utility and the core for games with incomplete information. *Journal of Economic Theory*, 136(1):260–285, 2007. DOI: 10.1016/j.jet.2006.08.002 Cited on page(s) 98, 111

[193] J. Nash. Two-person cooperative games. *Econometrica*, 21(1):128–140, 1953. DOI: 10.2307/1906951 Cited on page(s) 109

[194] N. Nisan. Introduction to mechanism design (for computer scientists). In N. Nisan, T. Roughgarden, E. Tardos, and V. Vazirani, editors, *Algorithmic Game Theory*, pages 209–242. Cambridge University Press, 2007. Cited on page(s) 111

[195] N. Nisan, T. Roughgarden, E. Tardos, and V. V. Vazirani, editors. *Algorithmic Game Theory*. Cambridge University Press, 2007. Cited on page(s) xv, 5, 10

[196] N. Ohta, V. Conitzer, R. Ichimura, Y. Sakurai, A. Iwasaki, and M. Yokoo. Coalition structure generation utilizing compact characteristic function representations. In *CP'09: 15th International Conference on Principles and Practice of Constraint Programming*, pages 623–638, 2009. DOI: 10.1007/978-3-642-04244-7-49 Cited on page(s) 87

[197] N. Ohta, V. Conitzer, Y. Satoh, A. Iwasaki, and M. Yokoo. Anonymity-proof Shapley value: extending Shapley value for coalitional games in open environments. In *AAMAS'08: 7th International Conference on Autonomous Agents and Multiagent Systems*, pages 927–934, 2008. Cited on page(s) 112

[198] N. Ohta, A. Iwasaki, M. Yokoo, K. Maruono, V. Conitzer, and T. Sandholm. A compact representation scheme for coalitional games in open anonymous environments. In *AAAI'06: 21st National Conference on Artificial Intelligence*, pages 697–702, 2006. Cited on page(s) 45, 46

[199] A. Okada. A noncooperative coalitional bargaining game with random proposers. *Games and Economic Behavior*, 16:97–108, 1996. DOI: 10.1006/game.1996.0076 Cited on page(s) 93

[200] M. J. Osborne and A. Rubinstein. *A Course in Game Theory*. MIT Press, 1994. Cited on page(s) 5, 7, 10, 83, 93, 110

[201] G. Owen. Multilinear extensions and the Banzhaf value. *Naval Research Logistics Quarterly*, 22(4):741–750, 1975. DOI: 10.1002/nav.3800220409 Cited on page(s) 57

[202] G. Owen. Values of games with a priori unions. In R. Henn and O. Moeschlin, editors, *Mathematical Economics and Game Theory*, pages 76–87. Springer-Verlag, 1977. Cited on page(s) 114

[203] C. H. Papadimitriou. *Computational Complexity*. Addison Wesley, 1994. Cited on page(s) xvi, 8

[204] C. H. Papadimitriou and K. Steiglitz. *Combinatorial Optimization*. Prentice Hall, 1982. Cited on page(s) 8, 30

[205] M. Pauly. A modal logic for coalitional power in games. *Journal of Logic and Computation*, 12(1):149–166, 2002. DOI: 10.1093/logcom/12.1.149 Cited on page(s) 114

[206] M. Pechoucek, V. Marik, and J. Barta. A knowledge-based approach to coalition formation. *IEEE Intelligent Systems*, 17:17–25, 2002. DOI: 10.1109/MIS.2002.1005627 Cited on page(s) 92

[207] D. Peleg and P. Sudhölter. *Introduction to the Theory of Cooperative Games*. Springer Verlag, 2007. Cited on page(s) 7, 10

[208] M. Perry and P. J. Reny. A noncooperative view of coalition formation and the core. *Econometrica*, 62(4):795–817, 1994. DOI: 10.2307/2951733 Cited on page(s) 93, 94

[209] J. A. M. Potters and S. Tijs. The nucleolus of a matrix game and other nucleoli. *Mathematics of Operations Research*, 17:164–174, 1992. DOI: 10.1287/moor.17.1.164 Cited on page(s) 109

[210] K. Prasad and J. S. Kelly. NP-completeness of some problems concerning voting games. *International Journal of Game Theory*, 19(1):1–9, 1990. DOI: 10.1007/BF01753703 Cited on page(s) 55

[211] A. Procaccia and J. Rosenschein. Learning to identify winning coalitions in the PAC model. In *AAMAS'06: 5th International Conference on Autonomous Agents and Multiagent Systems*, pages 673–675, 2006. DOI: 10.1145/1160633.1160751 Cited on page(s) 105

[212] D. Pudjianto, C. Ramsay, and G. Strbac. Virtual power plant and system integration of distributed energy resources. *IET Renewable Power Generation*, 1(1):10–16, 2007. DOI: 10.1049/iet-rpg:20060023 Cited on page(s) 116, 117

[213] M. L. Puterman. *Markov Decision Processes*. Wiley, 1994. Cited on page(s) 102

[214] T. Rahwan and N. R. Jennings. Coalition structure generation: Dynamic programming meets anytime optimisation. In *AAAI'08: 23rd AAAI Conference on Artificial Intelligence*, pages 156–161, 2008. Cited on page(s) 92

[215] T. Rahwan and N. R. Jennings. An improved dynamic programming algorithm for coalition structure generation. In *AAMAS'08: 7th International Conference on Autonomous Agents and Multiagent Systems*, pages 1417–1420, 2008. Cited on page(s) 92

[216] T. Rahwan, T. Michalak, N. R. Jennings, M. Wooldridge, and P. McBurney. Coalition structure generation in multi-agent systems with positive and negative externalities. In *IJCAI'09: 21st International Joint Conference on Artificial Intelligence*, pages 257–263, 2009. Cited on page(s) 85

[217] T. Rahwan, T. Michalak, N. R. Jennings, M. Wooldridge, and P. McBurney. Coalition structure generation in multi-agent systems with positive and negative externalities. In *IJCAI'09: 21st International Joint Conference on Artificial Intelligence*, pages 257–263, 2009. Cited on page(s) 92, 119

[218] T. Rahwan, S. D. Ramchurn, V. D. Dang, A. Giovannucci, and N. R. Jennings. Anytime optimal coalition structure generation. In *AAAI'07: 22nd AAAI Conference on Artificial Intelligence*, pages 1184–1190, 2007. Cited on page(s) 91, 92

[219] E. Resnick, Y. Bachrach, R. Meir, and J. S. Rosenschein. The cost of stability in network flow games. In *MFCS'09: 34th International Symposium on Mathematical Foundations of Computer Science*, pages 636–650, 2009. DOI: 10.1007/978-3-642-03816-7-54 Cited on page(s) 41

[220] M. H. Rothkopf, A. Pekec, and R. M. Harstad. Computationally manageable combinatorial auctions. *Management Science*, 44(8):1131–1147, 1995. DOI: 10.1287/mnsc.44.8.1131 Cited on page(s) 88

[221] A. Rubinstein. Perfect equilibrium in a bargaining model. *Econometrica*, 50(1):97–110, 1982. DOI: 10.2307/1912531 Cited on page(s) 93

[222] W. Saad, Z. Han, M. Debbah, and A. Hjørungnes. A distributed merge and split algorithm for fair cooperation in wireless networks. In *IEEE ICC Workshop on Cooperative Communications and Networking*, pages 311–315, 2008. DOI: 10.1109/ICCW.2008.65 Cited on page(s) 116

[223] W. Saad, Z. Han, M. Debbah, A. Hjørungnes, and T. Basar. Coalitional game theory for communication networks: A tutorial. *IEEE Signal Processing Magazine*, 26(5):77–97, 2009. DOI: 10.1109/MSP.2009.000000 Cited on page(s) 115

[224] T. Sandholm, K. Larson, M. Andersson, O. Shehory, and F. Tohme. Coalition structure generation with worst case guarantees. *Artificial Intelligence*, 111(1–2):209–238, 1999. DOI: 10.1016/S0004-3702(99)00036-3 Cited on page(s) 87, 89, 91

[225] D. Schmeidler. The nucleolus of a characteristic function game. *SIAM Journal on Applied Mathematics*, 17:1163–1170, 1969. DOI: 10.1137/0117107 Cited on page(s) 32, 33

[226] A. Schrijver. *Combinatorial Optimization: Polyhedra and Efficiency*. Springer, 2003. Cited on page(s) 30

[227] R. Serrano. Strategic bargaining, surplus sharing problems and the nucleolus. *Journal of Mathematical Economics*, 24:319–329, 1995. DOI: 10.1016/0304-4068(94)00696-8 Cited on page(s) 109

[228] R. Serrano. Fifty years of the Nash program, 1953–2003. *Investigaciones Económicas*, 29:219–258, 2005. Cited on page(s) 109

[229] R. Serrano and R. Vohra. Non-cooperative implementation of the core. *Social Choice and Welfare*, 14:513–525, 1997. DOI: 10.1007/s003550050084 Cited on page(s) 93, 94

[230] T. C. Service and J. A. Adams. Constant factor approximation algorithms for coalition structure generation. *Autonomous Agents and Multi-Agent Systems*, 23(1):1–17, 2011. DOI: 10.1007/s10458-010-9124-7 Cited on page(s) 92

[231] L. S. Shapley. A value for n-person games. In H. W. Kuhn and A. W. Tucker, editors, *Contributions to the Theory of Games, volume II*, pages 307–317. Princeton University Press, 1953. Cited on page(s) 17

[232] L. S. Shapley. On balanced sets and cores. *Naval Research Logistics Quarterly*, 14:453–460, 1967. DOI: 10.1002/nav.3800140404 Cited on page(s) 31

[233] L. S. Shapley. Cores of convex games. *International Journal of Game Theory*, 1(1):11–26, 1971. DOI: 10.1007/BF01753431 Cited on page(s) 27, 35

[234] L. S. Shapley and M. Shubik. The assignment game I: The core. *International Journal of Game Theory*, 1:111–130, 1972. DOI: 10.1007/BF01753437 Cited on page(s) 42

[235] O. Shehory and S. Kraus. Task allocation via coalition formation among autonomous agents. In *IJCAI'95: 14th International Joint Conference on Artificial Intelligence*, pages 655–661, 1995. Cited on page(s) 94

[236] O. Shehory and S. Kraus. Formation of overlapping coalitions for precedence-ordered task-execution among autonomous agents. In *ICMAS'96: 2nd International Conference on Multiagent Systems*, pages 330–337, 1996. DOI: 10.1007/s10846-007-9150-0 Cited on page(s) 113

[237] O. Shehory and S. Kraus. Methods for task allocation via agent coalition formation. *Artificial Intelligence*, 101(1–2):165–200, 1998. DOI: 10.1016/S0004-3702(98)00045-9 Cited on page(s) 96

[238] O. Shehory and S. Kraus. Feasible formation of coalitions among autonomous agents in nonsuperadditive environments. *Computational Intelligence*, 15:218–251, 1999. DOI: 10.1111/0824-7935.00092 Cited on page(s) 96

[239] O. Shehory, K. Sycara, and S. Jha. Multiagent coordination through coalition formation. In *ATAL'97: 4th International Workshop on Agent Theories, Architectures and Languages*, 1997. DOI: 10.1007/BFb0026756 Cited on page(s) 96

[240] Y. Shoham and K. Leyton-Brown. *Multiagent Systems: Algorithmic, Game-Theoretic, and Logical Foundations*. Cambridge University Press, 2008. Cited on page(s) xv, 5, 10

[241] T. Solymosi and T. E. S. Raghavan. An algorithm for finding the nucleolus of assignment games. *International Journal of Game Theory*, 23:119–143, 1994. DOI: 10.1007/BF01240179 Cited on page(s) 42

[242] J. Suijs and P. Borm. Stochastic cooperative games: superadditivity, convexity and certainty equivalents. *Games and Economic Behavior*, 27:331–345, 1999. DOI: 10.1006/game.1998.0672 Cited on page(s) 97, 98

[243] J. Suijs, P. Borm, A. De Wagenaere, and S. Tijs. Cooperative games with stochastic payoffs. *European Journal of Operational Research*, 113:193–205, 1999. DOI: 10.1016/S0377-2217(97)00421-9 Cited on page(s) 97, 98

[244] S. C. Sung and D. Dimitrov. On core membership testing for hedonic coalition formation games. *Operations Research Letters*, 35:155–158, 2007. DOI: 10.1016/j.orl.2006.03.011 Cited on page(s) 77, 78

[245] S. C. Sung and D. Dimitrov. On myopic stability concepts for hedonic games. *Theory and Decision*, 52(1):31–45, 2007. DOI: 10.1007/s11238-006-9022-2 Cited on page(s) 77

[246] S. C. Sung and D. Dimitrov. Computational complexity in additive hedonic games. *European Journal of Operational Research*, 203(3):635–639, 2010. DOI: 10.1016/j.ejor.2009.09.004 Cited on page(s) 77

[247] R. S. Sutton and A. G. Barto. *Reinforcement Learning: An Introduction*. MIT Press, 1998. Cited on page(s) 102

[248] K. Sycara, K. Decker, A. Pannu, M. Williamson, and D. Zeng. Distributed intelligent agents. *IEEE Expert-Intelligent Systems and Their Applications*, 11(6):36–45, 1996. DOI: 10.1109/64.546581 Cited on page(s) 95

[249] K. Sycara and D. Zeng. Coordination of multiple intelligent software agents. *International Journal of Intelligent and Cooperative Information Systems*, 5(2 & 3):181–211, 1996. Cited on page(s) 95

[250] A. D. Taylor. *Mathematics and Politics*. Springer Verlag, 1995. Cited on page(s) 66

[251] A. D. Taylor and W. S. Zwicker. *Simple games: Desirability Relations, Trading, Pseudoweightings*. Princeton University Press, 1999. Cited on page(s) 10, 65, 66, 67

[252] M. Tennenholtz. Program equilibrium. *Games and Economic Behavior*, 49:363–373, 2004. DOI: 10.1016/j.geb.2004.02.002 Cited on page(s) 109, 110

[253] U.S. Department of Energy. Grid 2030: A national vision for electricity's second 100 years, 2003. Cited on page(s) 116

[254] R. van den Brink. On hierarchies and communication. Tinbergen Discussion Paper 06/056-1, Tinbergen Institute and Free University, 2006. Cited on page(s) 119

[255] W. Vickrey. Counterspeculation, auctions, and competitive sealed tenders. *The Journal of Finance*, 16(1):8–37, 1961. DOI: 10.2307/2977633 Cited on page(s) 118

[256] M. Vidyasagar. *A Theory of Learning and Generalization: With Applications to Neural Networks and Control Systems*. Springer Verlag, 1997. Cited on page(s) 105

[257] J. von Neumann and O. Morgenstern. *Theory of Games and Economic Behavior*. Princeton University Press, 1944. Cited on page(s) 34

[258] P. Vytelingum, S. D. Ramchurn, T. D. Voice, A. Rogers, and N. R. Jennings. Trading agents for the smart electricity grid. In *AAMAS'10: 9th International Conference on Autonomous Agents and Multiagent Systems*, pages 897–904, 2010. Cited on page(s) 116

[259] P. Vytelingum, T. D. Voice, S. D. Ramchurn, A. Rogers, and N. R. Jennings. Agent-based micro-storage management for the smart grid. In *AAMAS'10: 9th International Conference on Autonomous Agents and Multiagent Systems*, pages 39–46, 2010. Cited on page(s) 116

[260] C. J. C. H. Watkins. *Learning from Delayed Rewards*. PhD thesis, University of Cambridge, England, 1989. Cited on page(s) 102

[261] M. Wooldridge. *An Introduction to MultiAgent Systems (2nd ed.)*. Wiley, 2009. Cited on page(s) xv, 5, 10

[262] M. Wooldridge, T. Ågotnes, P. E. Dunne, and W. van der Hoek. Logic for automated mechanism design—a progress report. In *AAAI'07: 22nd AAAI Conference on Artificial Intelligence*, pages 9–16, 2007. Cited on page(s) 115

[263] M. Wooldridge and P. E. Dunne. On the computational complexity of qualitative coalitional games. *Artificial Intelligence*, 158(1):27–73, 2004. DOI: 10.1016/j.artint.2004.04.002 Cited on page(s) 79, 80

[264] M. Wooldridge and P. E. Dunne. On the computational complexity of coalitional resource games. *Artificial Intelligence*, 170(10):835–871, 2006. DOI: 10.1016/j.artint.2006.03.003 Cited on page(s) 81, 82

[265] J. Yamamoto and K. Sycara. A stable and efficient buyer coalition formation scheme for e-marketplaces. In *Agents'01: 5th International Conference on Autonomous Agents*, pages 576–583, Montreal, Canada, 2001. DOI: 10.1145/375735.376452 Cited on page(s) 111

[266] H. Yan. Noncooperative selection of the core. *International Journal of Game Theory*, 31(4):527–540, 2003. DOI: 10.1007/s001820300137 Cited on page(s) 93, 94

[267] D. Y. Yeh. A dynamic programming approach to the complete set partitioning problem. *BIT Numerical Mathematics*, 26(4):467–474, 1986. DOI: 10.1007/BF01935053 Cited on page(s) 88

[268] C. S. K. Yeung, A. S. Y. Poon, and F. F. Wu. Game theoretical multi-agent modeling of coalition formation for multirateral trade. *IEEE Transactions on Power Systems*, 14(3):929–934, 1999. DOI: 10.1109/59.780905 Cited on page(s) 116

[269] M. Yokoo, V. Conitzer, T. Sandholm, N. Ohta, and A. Iwasaki. Coalitional games in open anonymous environments. In *AAAI'05: 20th National Conference on Artificial Intelligence*, pages 509–514, 2005. Cited on page(s) 112

[270] Y. Zick and E. Elkind. Arbitrators in overlapping coalition formation games. In *AAMAS'11: 10th International Conference on Autonomous Agents and Multiagent Systems*, pages 55–62, 2011. Cited on page(s) 114, 119

[271] Y. Zick, A. Skopalik, and E. Elkind. The Shapley value as a function of the quota in weighted voting games. In *IJCAI'11: 21st International Joint Conference on Artificial Intelligence*, pages 490–496, 2011. Cited on page(s) 53, 55

[272] M. Zuckerman, P. Faliszewski, Y. Bachrach, and E. Elkind. Manipulating the quota in weighted voting games. In *AAAI'08: 23rd AAAI Conference on Artificial Intelligence*, pages 215–220, 2008. Cited on page(s) 53

Authors' Biographies

GEORGIOS CHALKIADAKIS

Georgios Chalkiadakis Georgios Chalkiadakis is an Assistant Professor at the Department of Electronic and Computer Engineering, Technical University of Crete (TUC). His research interests are in the areas of multiagent systems, algorithmic game theory, and learning, and more specifically, in coalition formation, decision making under uncertainty, and reinforcement learning in multiagent domains. His PhD thesis (University of Toronto, 2007) was nominated for the IFAAMAS-2007 Victor Lesser Distinguished Dissertation Award. Georgios has served as Program Committee member for several top international conferences, and as a reviewer for several top journals in his areas of expertise. Before joining TUC, he was a Research Fellow at the School of Electronics and Computer Science, University of Southampton, UK; and has also worked as a software engineer at the Institute of Computer Science of the Foundation for Research and Technology - Hellas, and as a teacher of informatics in Greek high schools.

EDITH ELKIND

Edith Elkind is an Assistant Professor and the National Research Foundation fellow at the Division of Mathematics, School of Physical and Mathematical Sciences, Nanyang Technological University (Singapore). Edith has an undergraduate degree in Mathematics from Moscow State University (Russia) and a PhD in Computer Science from Princeton University (2005). Prior to joining NTU, she worked as a postdoctoral researcher at University of Warwick (UK), University of Liverpool (UK) and Hebrew University of Jerusalem (Israel), as well as a lecturer at University of Southampton (UK). Her research interests are in the area of algorithmic game theory and computational social choice, and she has published over 50 papers in top international conferences and journals in these fields. Edith has served as a senior PC member of major algorithmic game theory and AI conferences, such as ACM EC, AAAI, IJCAI and AAMAS. She is an editorial board member of Journal of Artificial Intelligence Research (JAIR), Journal of Autonomous Agents and Multiagent Systems (JAAMAS) and ACM Transactions on Economics and Computation (TEAC).

MICHAEL WOOLDRIDGE

Michael Wooldridge is a Professor in the Department of Computer Science at the University of Liverpool, UK. He has been active in multi-agent systems research since 1989, and has published over three hundred articles in the area. His main interests are in the use of formal methods for reasoning about autonomous agents and multi-agent systems. Wooldridge was the recipient of the ACM Autonomous Agents Research Award in 2006. He is an associate editor of the journals "Artificial Intelligence" and "Journal of AI Research (JAIR)". His introductory textbook "An Introduction to Multiagent Systems" was published by Wiley in 2002 (second edition 2009).

Index

Printed in the United States
by Baker & Taylor Publisher Services